1st Grade Math Workbook

Addition and Subtraction

Want Free Extra Goodies for Your Student?
Email us at: info@homerunpress.com.
Title the email "1st Grade Math Workbook"
and we'll send some extra worksheets your way!

We create our workbooks with love and great care.
For any issues with your workbook, such as printing errors, typos, faulty binding, or something else, please do not hesitate to contact us at: info@homerunpress.com. We will make sure you get a replacement copy immediately.

THANK YOU!

Copyright © 2020 Home Run Press, LLC. All rights reserved. Registered in U.S. Patent and Trademark Office.

Home Run Press, LLC 1603 Capitol Ave. Suite 310 A551 Cheyenne, WY 82001 USA
www.homerunpress.com

The purchase of this material entitles the buyer to reproduce worksheets and activities for classroom use only – not for commercial resale. Reproduction of these materials for an entire school or district is strictly prohibited. No part of this book may be reproduced (except as noted before), stored in retrieval system, or transmitted in any form or by any means (mechanically, electronically, photocopying, recording, etc.) without the prior written consent of Home Run Press, LLC.

First published in the USA 2020.   ISBN 9781952368042

Table of Contents

Understanding addition. A Number Line. Maze	5-10
Adding without Regrouping. Number Bonds	11-17
Understanding Subtraction. A Number Line	18-21
Subtracting without Regrouping. Number Bonds	22-28
Place Value	29-30
Adding 2-Digit Numbers. Number Bonds. Maze	31-38
Subtracting 2-Digit Numbers. Number Bonds. Word Problems	39-45
Adding 3-Digit Numbers. Place Value. Number Series	46-51
Subtracting 3-Digit Numbers. Number Bonds. Place value. Puzzles	52-53
Comparing	54-55
Rounding	56-58
Even and Odd Numbers. Estimation	59-60
Money	61-63
Scales Balance	64-65
Understanding Fractions	66-69
Problem-Solving: Reasoning	70-73
Word Problems	74-76
Adding with Regrouping. CogAT test prep	77-81
Subtracting with Regrouping. CogAT test prep	82-84
Adding and Subtracting with Regrouping: Timed Tests	85-94
Answers	95

Hi. I'm Sunny. For me, everything is an adventure. I am ready to try anything, take chances, see what happens - and help you try, too! I like to think I'm confident, caring and have an open mind. I will cheer for your success and encourage everyone! I'm ready to be a really good friend!

I've got a problem. Well, I've always got a problem. And I don't like it. It makes me cranky, and grumpy, impatient and the truth is, I got a bad attitude. There. I said it. I admit it. And the reason I feel this way? Math! I don't get it and it bums me out. Grrrr!

Not trying to brag, but I am the smartest Brainer that ever lived - and I'm a brilliant shade of blue. That's why they call me Smarty. I love to solve problems and I'm always happy to explain how things work - to help any Brainer out there! To me, work is fun, and math is a blast!

I scare easily. Like, even just a little …Boo! Oh wow, I've scared myself! Anyway, they call me Pickles because I turn a little green when I get panicky. Especially with new stuff. Eek! And big complicated problems. Really any problem. Eek! There, I did it again.

Hi! Name's Pepper. I have what you call a positive outlook. I just think being alive is exciting! And you know something? By being friendly, kind and maybe even wise, you can have a pretty awesome day every day on this amazing planet.

A famous movie star once said, "I want to be alone." Well, I do too! I'm best when I'm dreaming, thinking, and in my own world. And so, I resist! Yes, I resist anything new, and only do things my way or quit. The rest of the Brainers have math, but I'd rather have a headache and complain. Or pout.

1. Read.

When I add 3 candies and 2 candies, there are 5 candies altogether. It does not matter which way I add candies together.

3 + 2 = 5 candies

means equals

means add or plus

2 + 3 = 5 candies

I add 4 cars and 3 trucks.

I have 4 cars and 3 trucks together. I can find the total simply by counting them all. There are 7 in all.

4 + 3 = 7

1. Read.

I use a number line to find out the answer when I add 4 and 2. First, I draw a line and mark it with numbers. I find 4 on the number line.

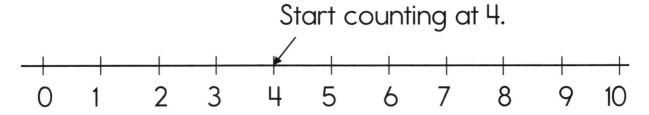

Start counting at 4.

I need to add 2, so I jump 2 places to the right.

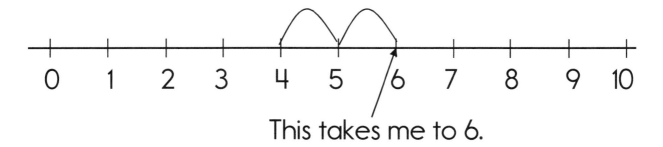

This takes me to 6.

So   4 + 2 = 6

I add 30 and 50. First, I find 30. Then, I jump 5 places to the right.

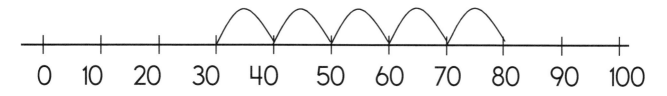

So   30 + 50 = 80

1. <u>Add</u>. Use a number line to show the jumps.

2 + 7 = ___
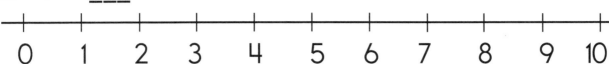

6 + 4 = ___
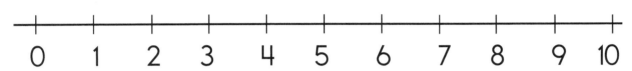

9 + 1 = ___
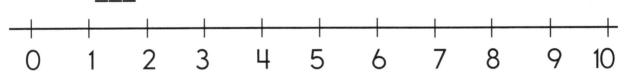

50 + 50 = ___
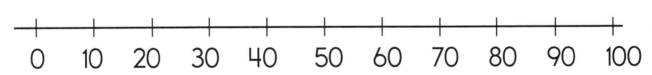

40 + 40 = ___
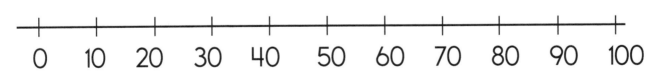

30 + 70 = ___
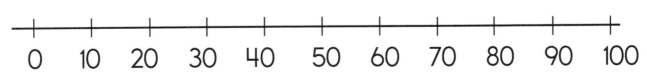

1. <u>Compare</u>, using ">," "<," or "=." The first one is done for you.

3 + 7	>	10 – 2	8 – 5		4 – 1
4 + 2		9 – 2	7 - 2		10 – 6
2 + 5		10 – 2	8 – 5		10 – 7
6 + 3		4 + 3	6 – 4		5 – 2

"We need to add 3, 4 and 2". <u>What</u> is the total?

2 + 3 + 4 = _

3 + 4 + 2 = _

4 + 3 + 2 = _

The order of addends does not affect the total – the sum will be the same.

Aha, I remember something about that. I've read that rule called the COMMUTATIVE property.

That's true. 2 candies + 3 candies or 3 candies + 2 candies = always 5 candies!

2. <u>Solve</u> the puzzle.

 + + + = 10

 =

 + + = 7

 =

1. <u>Add</u> and <u>change</u> the addends' order. The first one is done for you.

2 + 5 + 3 = 10   5 + 3 + 2 = 10   3 + 2 + 5 = 10

1 + 4 + 2 = _   _ + _ + _ = _   _ + _ + _ = _

5 + 0 + 5 = _   _ + _ + _ = _   _ + _ + _ = _

4 + 1 + 3 = _   _ + _ + _ = _   _ + _ + _ = _

6 + 1 + 3 = _   _ + _ + _ = _   _ + _ + _ = _

2. <u>Complete</u> an addition number sentence with tens and ones.

16 = 10 + 6   18 = __ + __   15 = __ + __

10 = __ + __   13 = __ + __   17 = __ + __

3. <u>Write</u> the missing numbers to make the comparison true.

1 + 9   =   _ − 4   9 − 4   =   _ − 2

3 + 4   =   10 − _   10 − _   =   1 + 3

4 + 4   =   _ + 2   _ − 5   =   8 − 7

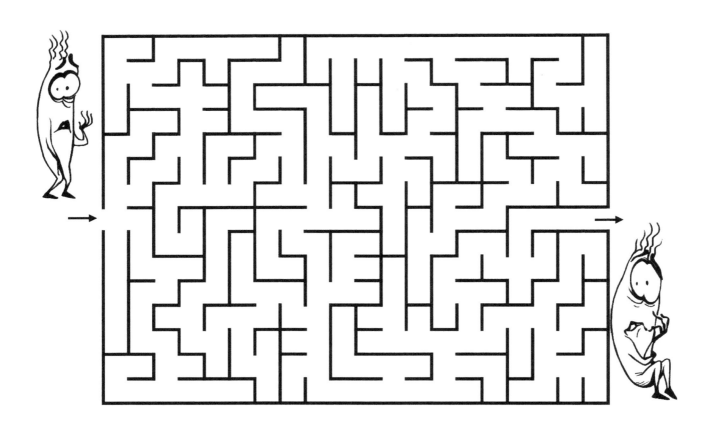

1.

I'm faster than Pickles, but Sunny is faster than me. Who is the fastest?

Answer:
_____

1. The blocks in each tower tells you how many hundreds, tens, and ones in each number. Write and put the numbers in order from the least to the greatest.

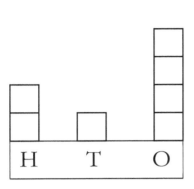 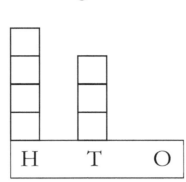 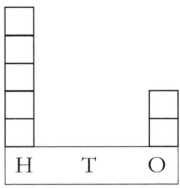

_____   _____   _____

2. Write and put the numbers in order from the largest to the smallest.

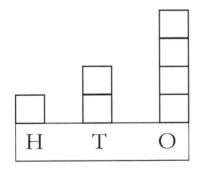 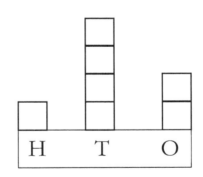 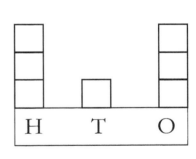

_____   _____   _____

3. My sister found 9 apples. She ate 6 apples. How many apples are left?

Answer: _____

1. Make and write the smallest and the biggest two-digit numbers you can with any two of these digits: 1, 7, 4, 6.

   Answer: _____

2. Make and write the smallest and the biggest two-digit numbers you can with any two of these digits: 9, 5, 9, 8.

   Answer: _____

3. Make and write the smallest and the biggest two-digit numbers you can with any two of these digits: 2, 9, 3, 0.

   Answer: _____

4. Write the missing number.

___ + 40 + 8 = 548   100 + ___ + 6 = 126

200 + 50 + ___ = 251   ___ + 80 + 9 = 389

___ + 10 + 0 = 610   400 + ___ + 4 = 484

700 + 30 + ___ = 737   ___ + 60 + 2 = 862

___ + 10 + 1 = 911   500 + ___ + 1 = 541

300 + 60 + ___ = 369   ___ + 70 + 2 = 572

1. <u>Read.</u>

I like to split the adding numbers into numbers that are easier to work with. I can show my favorite strategy. T = tens, O = ones.

Step 1. Let's add 12 and 15.

```
T O   T O
12 + 15 = ___
```

Step 2. Add the tens together.

```
T O   T O   T O
10 + 10 = 20
```

Step 3. Add the ones together.

```
T O   T O   T O
 2 +  5 =  7
```

Step 4. Add the tens and ones to find the total.

```
T    O   T O
20 + 7 = 27
```

1. <u>Add.</u>

16 + 12 = 10 + 10 + 6 + 2 = ___ + ___ = ___

21 + 17 = ___ + ___ + ___ + ___ = ___ + ___ = ___

13 + 15 = ___ + ___ + ___ + ___ = ___ + ___ = ___

I can add numbers using column addition. Hint: Write ones under ones. Write tens under tens.

Step 1: Write the digits that have the same place value lined up one above the other.

```
tens  ones
  1    2
+ 1    5
  ─    ─
```

Step 2: Start by adding the ones together. Add 2 ones and 5 ones: 2 + 5 = 7
Write 7 in the ones column.

```
tens  ones
  1   (2)
+ 1   (5)
  ─   (7)
```

Step 3: Add 1 ten and 1 ten. But I actually add 10 and 10. So the answer is: 10 + 10 = 20
I write 2 in the tens column.

```
 tens  ones
 (1)    2
+(1)    5
 (2)    7
```

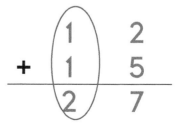

1. <u>Add.</u>

```
   3        5        3        7        4        8
+  3     +  2     +  6     +  2     +  4     +  1
───        ───      ───      ───      ───      ───
   6
```

```
   2        5        3        4        1        6
+  8     +  5     +  7     +  6     +  9     +  4
───        ───      ───      ───      ───      ───
```

2. <u>Complete</u> each pair of number bonds.

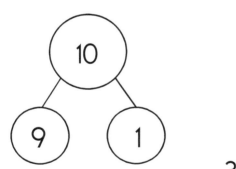

9 and 1 make 10   3 and __ make 10

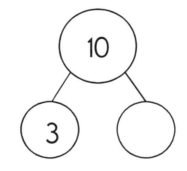

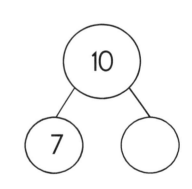

7 and __ make 10

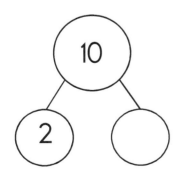

2 and __ make 10

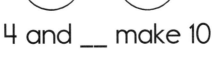

4 and __ make 10   5 and __ make 10

1. <u>Use</u> cupcakes to make 10. <u>Color</u> the cupcakes brown and yellow.

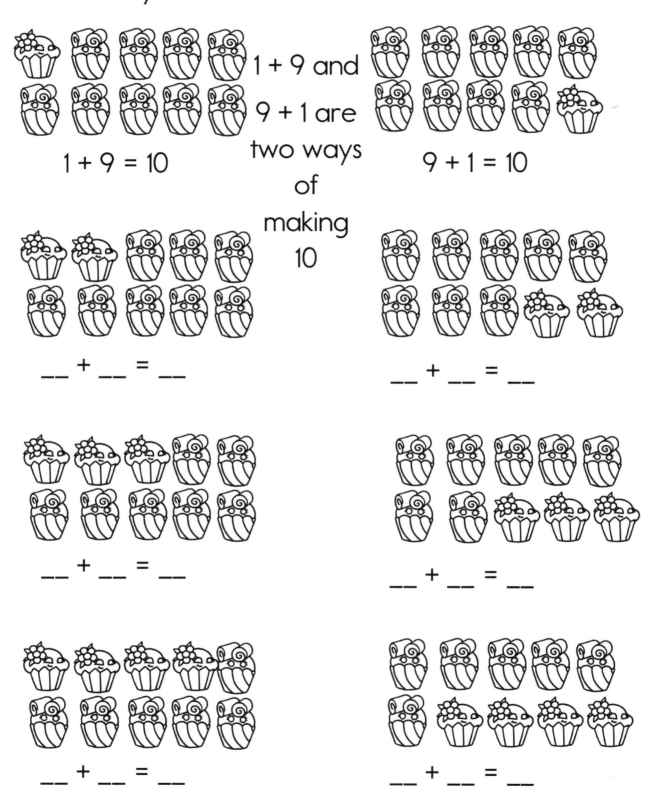

1. <u>Complete</u> each picture. When a shape is symmetrical, each half is a mirror image of the other.

Symmetry

line of symmetry

2. <u>Add.</u>

```
   1      2      3      6      4      8
 + 7    + 5    + 3    + 2    + 1    + 1
 ———    ———    ———    ———    ———    ———
   8
```

```
   2      1      3      2      3      6
 + 6    + 5    + 5    + 2    + 4    + 1
 ———    ———    ———    ———    ———    ———
```

Subtraction is the opposite of addition. Subtraction means finding the difference between two numbers or taking away from a number. When I give 2 candies to my sister out of 3 candies that I have, how many candies are left?

means equals   means subtract or minus

3 - 2 = 1 candy

When I subtract or take away 2 cars from the 4 cars that my brother has, he is left with 2 cars.

He has 4 cars and I take away 2 cars. I can find the total simply by crossing out the 2 cars from the 4 cars. There are 2 cars left. 4 - 2 = 2

1. Read.

I use a number line to find out the answer when I subtract 4 from 7. First, I draw a line and mark it with numbers. I find 7 on the number line.

Start counting at 7.

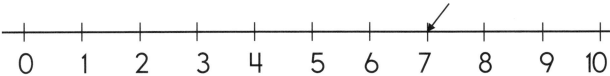

I need to take away 4, so I jump 4 places to the left.

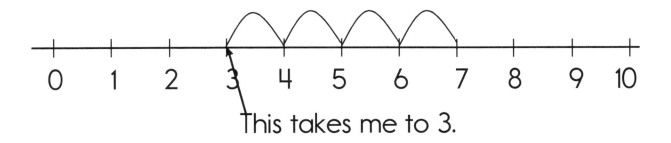
This takes me to 3.

So   7 - 4 = 3

I subtract 40 from 60. First, I find 60. Then, I jump 4 places to the left.

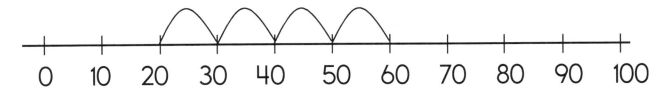

So   60 - 40 = 20

1. <u>Subtract</u>. Use a number line to show the jumps.

9 - 6 = ___

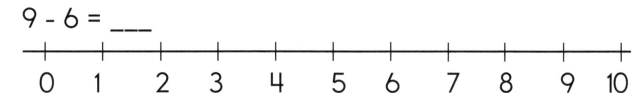

7 - 5 = ___

10 - 7 = ___

80 - 30 = ___

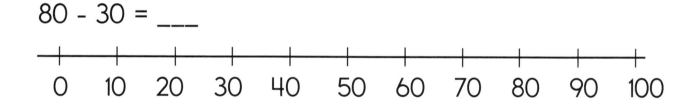

60 - 50 = ___

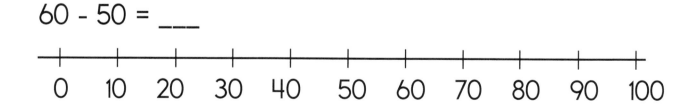

100 - 70 = ___

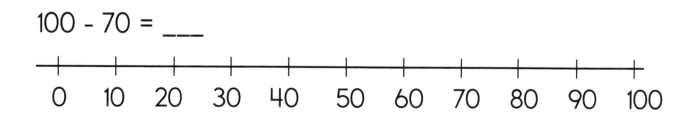

1. Read.

I like to split the numbers I subtract into numbers that are easier to work with. I can show my favorite strategy.

Step 1. Let's subtract 13 from 38.

T O	T O
3 8 - 1 3 = ___	

Step 2. Subtract the tens from 38.

T O	T O	T O
3 8 - 1 0 = 2 8		

Step 3. Subtract the ones from the remaining 28.

T O	T O	T O
2 8 - 3 = 2 5		

2. Subtract.

5 – 1 = ___   6 – 1 = ___   9 – 1 = ___

8 – 2 = ___   4 – 2 = ___   7 – 2 = ___

3 – 3 = ___   6 – 3 = ___   8 – 3 = ___

7 – 4 = ___   9 – 4 = ___   5 – 4 = ___

I can subtract numbers using column subtraction. Hint: Write ones under ones. Write tens under tens.

Step 1: Write the digits that have the same place value lined up one above the other.

tens	ones
2	9
− 1	3
−	−

Step 2: Subtract 3 ones from 9 ones: 9 − 3 = 6
Write 6 in the ones column.

tens	ones
2	(9)
− 1	(3)
−	(6)

Step 3: Subtract 1 ten from 2 tens:
2 − 1 = 1
I write 1 in the tens column.

tens	ones
(2)	9
− (1)	3
(1)	6

1. <u>Subtract.</u>

```
   9      5      8      7      4      8
 - 3    - 2    - 4    - 2    - 4    - 1
 ───    ───    ───    ───    ───    ───
   6
```

```
   8      5      5      7      9      6
 - 2    - 5    - 2    - 6    - 1    - 4
 ───    ───    ───    ───    ───    ───
```

2. <u>Complete</u> each pair of number bonds.

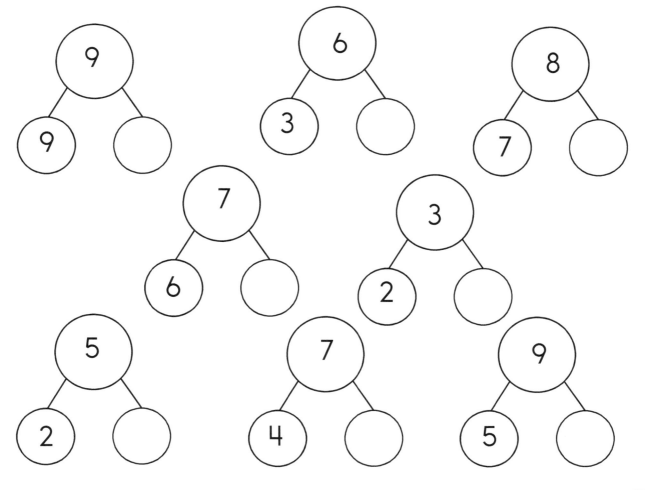

1. Subtract.

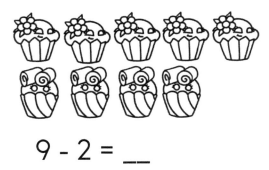

9 - 2 = __

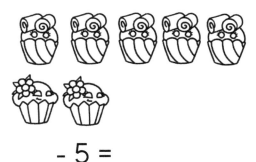

__ - 5 = __

__ - 4 = __

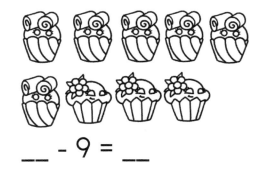

__ - 9 = __

__ - 3 = __

__ - 4 = __

__ - 6 = __

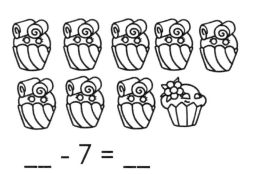

__ - 7 = __

1. If 5 cupcakes are eaten, how many are left?

10 - 5 = ___

2. If 7 cupcakes are eaten, how many are left?

___ - ___ = ___

3. If 4 cupcakes are eaten, how many are left?

___ - ___ = ___

4. If 9 cupcakes are eaten, how many are left?

___ - ___ = ___

1. <u>Subtract.</u>

___ - 5 = ___

___ - 7 = ___

___ - 4 = ___

___ - 2 = ___

2. <u>Subtract.</u>

```
  9      8      6      6      4      8
- 7    - 5    - 3    - 1    - 3    - 4
---    ---    ---    ---    ---    ---
  2
```

```
  7      7      8      2      6      6
- 4    - 5    - 3    - 2    - 4    - 5
---    ---    ---    ---    ---    ---
```

1. <u>Color</u> the flowers with the smaller number in each group.

1. Subtract.

```
  10      10      10      10      10      10
-  3    -  5    -  6    -  2    -  4    -  1
───     ───     ───     ───     ───     ───
   7
```

```
  17      18      13      15      16      19
-  6    -  5    -  1    -  2    -  4    -  1
───     ───     ───     ───     ───     ───
```

2. Complete each pair of number bonds.

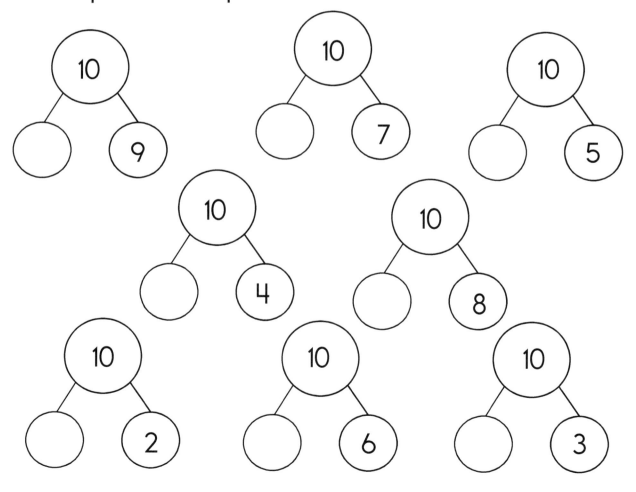

1. Read. Write how many tens. Write the number.

Two digit numbers are made up of tens and ones.

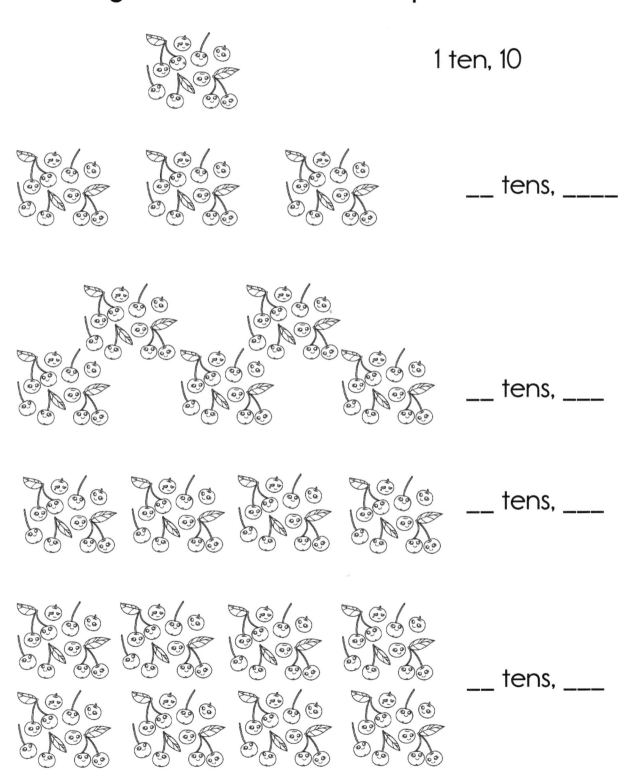

1 ten, 10

__ tens, ____

__ tens, ___

__ tens, ___

__ tens, ___

1. How many are there in each group?

10 + 1 makes 11

__ + __ makes __

__ + __ makes __

__ + __ makes __

__ + __ makes __

1. <u>Add.</u>

```
  10      10      10      10      10      10
+  3    +  5    +  6    +  2    +  4    +  1
----    ----    ----    ----    ----    ----
  13
```

```
  12      11      13      12      13      16
+  6    +  5    +  5    +  2    +  4    +  1
----    ----    ----    ----    ----    ----
```

2. <u>Complete</u> each pair of number bonds.

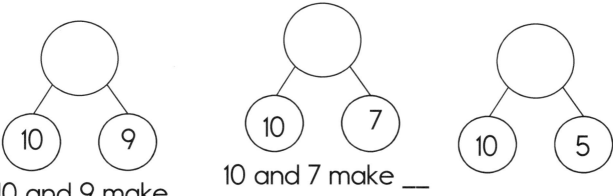

10 and 9 make __

10 and 7 make __

10 and 5 make __

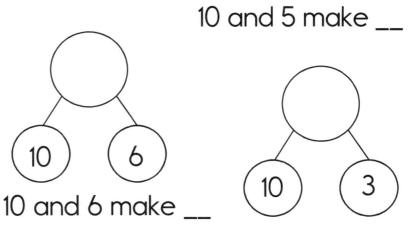

10 and 2 make __

10 and 6 make __

10 and 3 make __

1. Add.

```
   1 1        3 2        2 3        8 1        5 4
 + 5        + 6        + 1        + 7        + 2
 ─────      ─────      ─────      ─────      ─────
   1 6        _ _        _ _        _ _        _ _


   3 1        5 2        9 1        4 5        4 0
 + 4        + 2        + 4        + 1        + 8
 ─────      ─────      ─────      ─────      ─────
   _ _        _ _        _ _        _ _        _ _


   2 5        3 3        6 1        1 2        6 7
 + 2 2      + 3 6      + 1 5      + 4 3      + 1 2
 ─────      ─────      ─────      ─────      ─────
   _ _        _ _        _ _        _ _        _ _
```

1. Subtract.

```
   1 9        8 7        4 8        3 5        6 9
 - 4        - 2        - 3        - 1        - 3
 ─────      ─────      ─────      ─────      ─────
   1 5        _ _        _ _        _ _        _ _


   5 9        6 9        7 6        9 4        2 8
 - 2 1      - 4 5      - 3 4      - 5 3      - 1 8
 ─────      ─────      ─────      ─────      ─────
   _ _        _ _        _ _        _ _        _ _


   3 5        8 7        9 8        9 8        7 7
 - 1 1      - 3 3      - 3 4      - 5 3      - 1 5
 ─────      ─────      ─────      ─────      ─────
   _ _        _ _        _ _        _ _        _ _
```

1. <u>Circle</u> the missing number from the choice box to make the inequality true.

20 < ___ < 31	a) 18	b) 32	c) 26
9 < ___ < 11	a) 15	b) 10	c) 7
5 < ___ < 15	a) 11	b) 20	c) 4
46 < ___ < 50	a) 38	b) 48	c) 58

2. <u>What</u> is the value of the 3 in each of these numbers? <u>Circle</u> the right answer.

613	a) Hundreds	b) Tens	c) Ones
352	a) Hundreds	b) Tens	c) Ones
134	a) Hundreds	b) Tens	c) Ones
943	a) Hundreds	b) Tens	c) Ones

3. I saw 8 frogs. 5 left. <u>How many frogs stayed?</u>

Answer: _____

1. **How many** are there in each group?

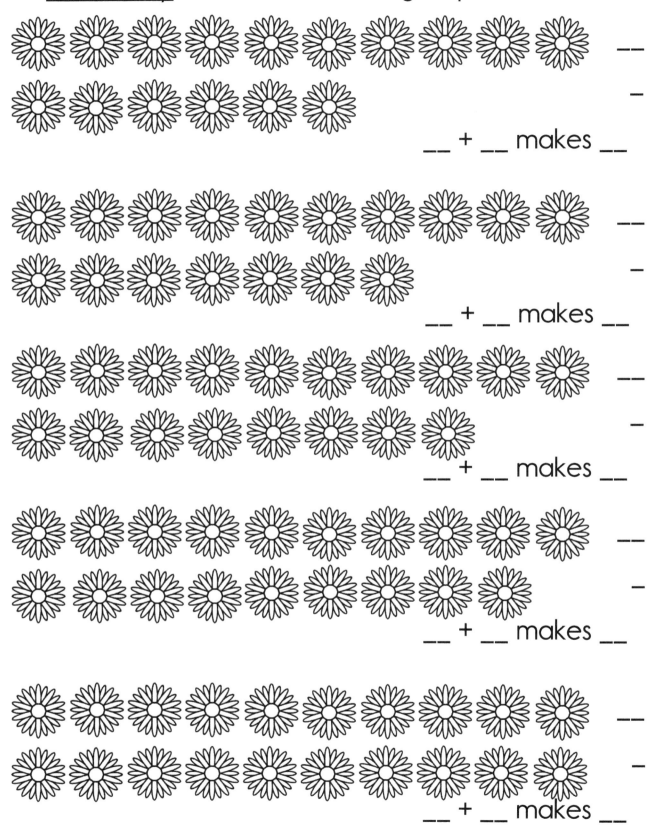

__ + __ makes __

__ + __ makes __

__ + __ makes __

__ + __ makes __

__ + __ makes __

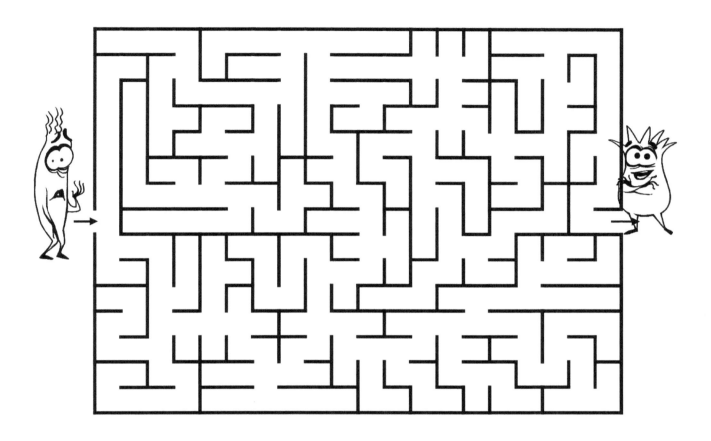

1. <u>Solve</u> the problem and <u>circle</u> ">," "<," or "=:"

1 cupcake = 10 candies.

3 cupcakes   > / < / =   2 cupcakes + 5 candies

Answer: _____.

1. <u>Add.</u>

```
  20      20      30      30      30      40
+  3    +  5    +  6    +  2    +  4    +  1
----    ----    ----    ----    ----    ----
  23
```

```
  42      51      53      62      63      66
+  6    +  5    +  5    +  2    +  4    +  1
----    ----    ----    ----    ----    ----
```

2. <u>Complete</u> each pair of number bonds.

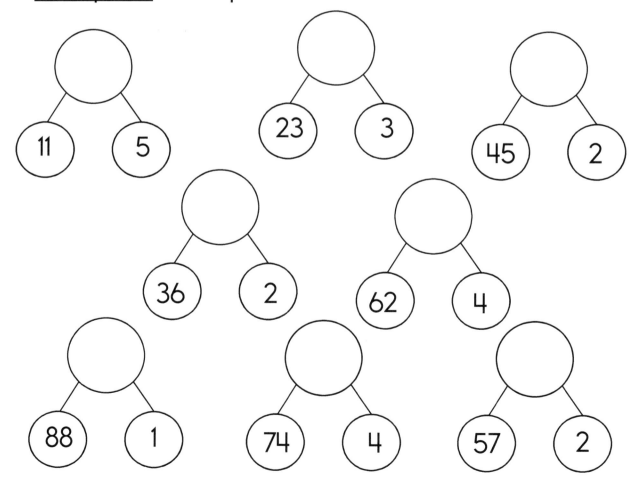

1. The blocks in each tower tell you how many hundreds, tens, and ones in each number. Write and put the numbers in order from the least to the

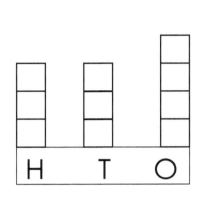 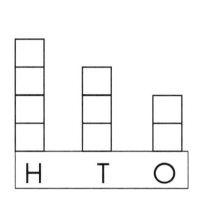 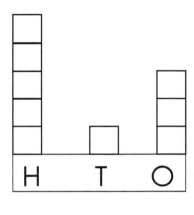

_____   _____   _____

2. Write the number for these.

2 tens + 5 ones = 25   1 ten + 9 ones = _____

3 tens + 0 ones = _____   5 tens + 6 ones = _____

4 tens + 9 ones = _____   6 tens + 2 ones = _____

7 tens + 0 ones = _____   8 tens + 5 ones = _____

3. What is the value of the digit 7 in the numbers below?

37   71   17   7   79

1. Add.

```
   11        14        15        17        13        12
+  12     + 14      + 21      + 31      + 14      + 16
  ----     ----     ----      ----      ----      ----
   23
```

```
   22        41        33        82        73        56
+  12     + 41      + 25      + 14      + 26      + 13
  ----     ----     ----      ----      ----      ----
```

2. Complete each pair of number bonds.

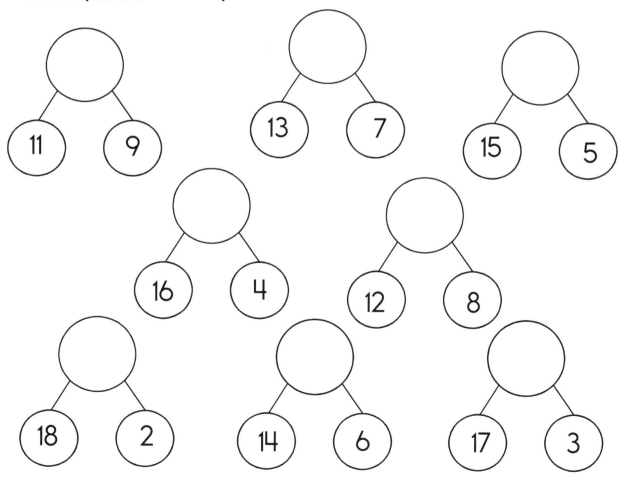

1. <u>Subtract.</u>

```
  25      28      37      33      37      42
-  3    -  5    -  6    -  2    -  4    -  1
----    ----    ----    ----    ----    ----
  22
```

```
  47      56      58      67      69      66
-  6    -  5    -  5    -  2    -  4    -  5
----    ----    ----    ----    ----    ----
```

2. <u>Complete</u> each pair of number bonds.

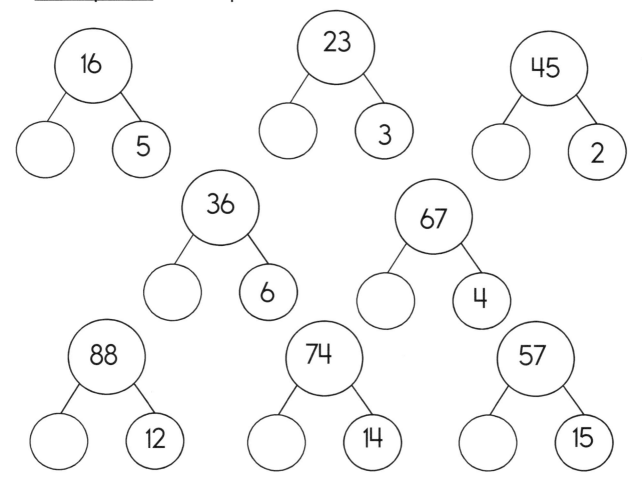

1. Subtract.

```
  34      28      45      37      57      47
-  12    - 14    - 21    - 31    - 14    - 16
-----   -----   -----   -----   -----   -----
  22


  22      64      57      39      48      33
-  12    - 41    - 25    - 14    - 26    - 11
-----   -----   -----   -----   -----   -----
```

2. Complete each pair of number bonds.

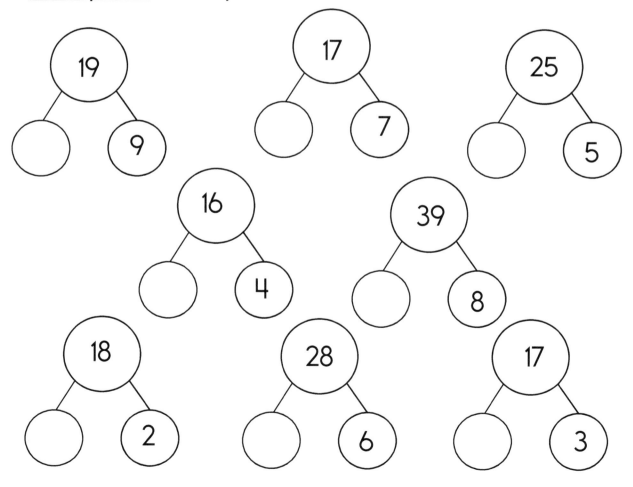

1. How many more do I need to add to the second group to make each group the same?

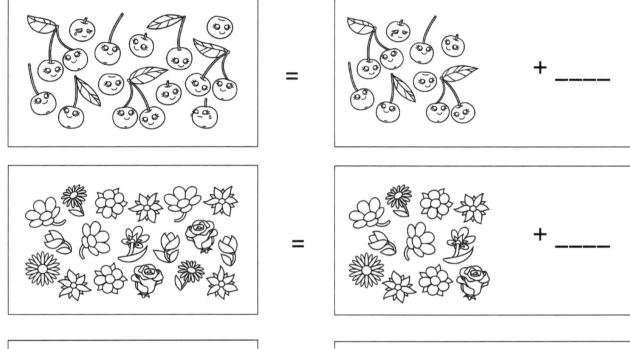

2. Circle all the combinations that equal 5.

10 – 5   16 – 11   8 – 2

   14 – 12   21 – 10   29 – 24

38 – 28   17 – 12   46 - 40

1. If 20 pencils are broken, how many are left?

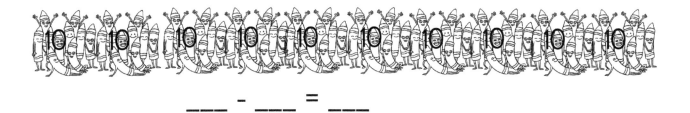

___ - ___ = ___

2. If 30 pencils are broken, how many are left?

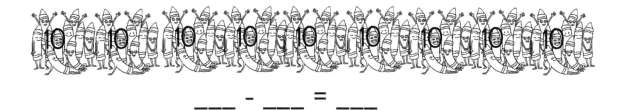

___ - ___ = ___

3. If 50 pencils are broken, how many are left?

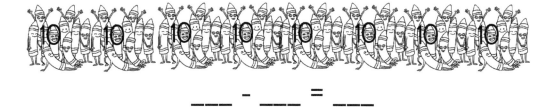

___ - ___ = ___

4. If 40 pencils are broken, how many are left?

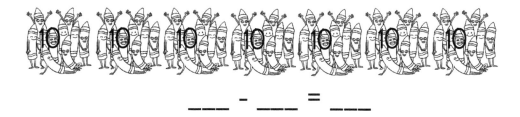

___ - ___ = ___

1. Venn Diagram: helps you sort things according to their different features.

I have many cards. ⬚3 of them are dragon-type cards. ⬚5 of them are flying-type cards. ⬚2 of them are both dragon- and flying-type cards. ⬚2 of them are neither dragon- nor flying-type cards. How many cards are there? Fill in the diagram.

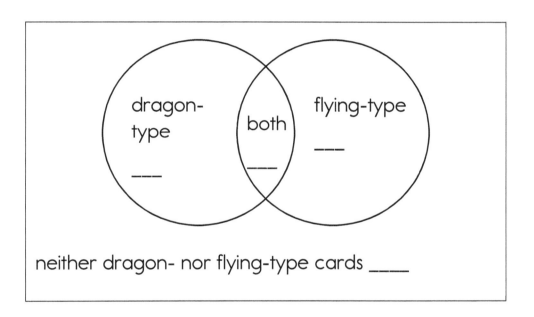

Answer: _____

2. Circle the right answer.

__ 2 + 3 6 = 58

a) 1
b) 2
c) 3

1. <u>Solve</u> the problem and <u>write</u> the missing number:

1 sunflower = 3 tulips

3 sunflowers = ___ tulips

Answer: _____.

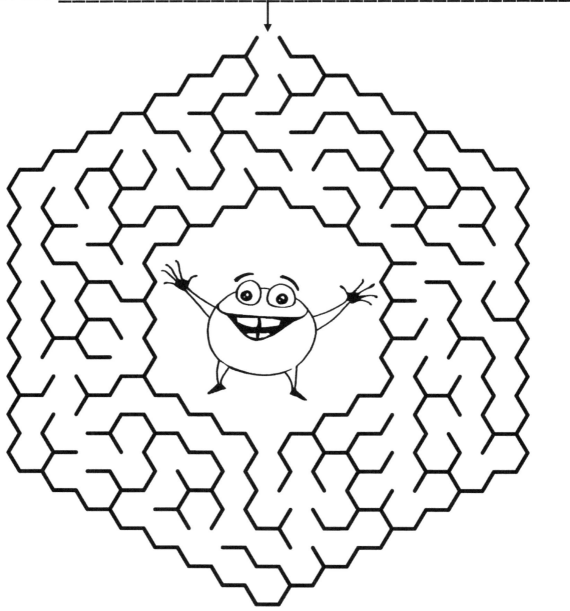

1. <u>Read</u>. <u>Write</u> how many tens. <u>Write</u> the number.

Three digit numbers are made up of hundreds, tens, and ones.

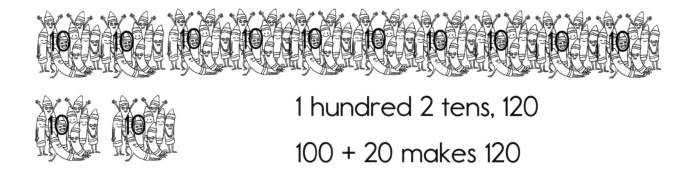

1 hundred 2 tens, 120

100 + 20 makes 120

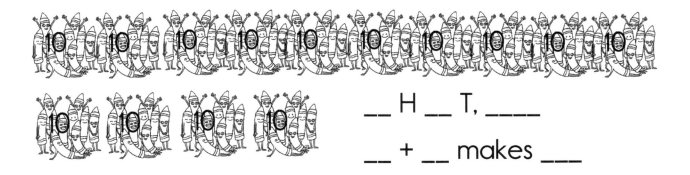

__ H __ T, ____

__ + __ makes ___

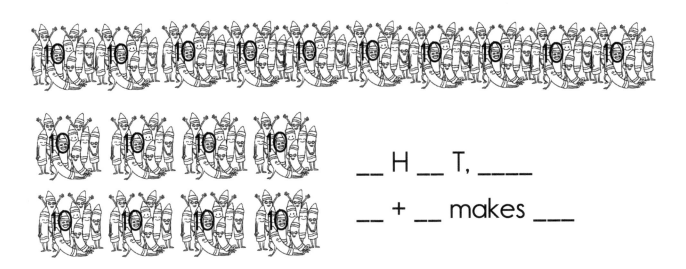

__ H __ T, ____

__ + __ makes ___

1. Add.

```
  100      100      100      100      100      100
+  25    +  15    +  46    +  12    +  47    +  71
-----    -----    -----    -----    -----    -----
  125
```

```
  120      110      130      120      130      160
+   6    +   5    +   5    +   2    +   4    +   1
-----    -----    -----    -----    -----    -----
```

2. Complete each pair of number bonds.

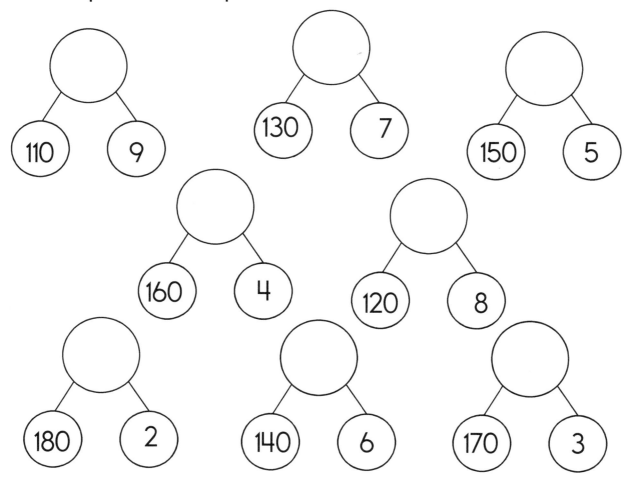

1. I am building a solid slab of rocks: the two rocks next to each other are added to get the number up above. Fill in the missing numbers.

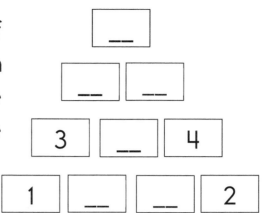

2. Write the numbers in order, from the smallest to the largest.

87, 9, 36, 14, 20, 71, 59, 23, 18, 65

_____

3. Complements to 20. Circle the missing numbers from the choice box to make the equations true.

8 + ___ = 20   a) 8   b) 2   c) 12

11 + ___ = 20   a) 10   b) 8   c) 9

5 + ___ = 20   a) 5   b) 15   c) 10

7 + ___ = 20   a) 13   b) 23   c) 3

2. Circle the right answer.

__ 6 + 2 3 = 99

a) 5
b) 9
c) 7

1. <u>Read.</u>

Place value is the amount a digit is worth in a number.

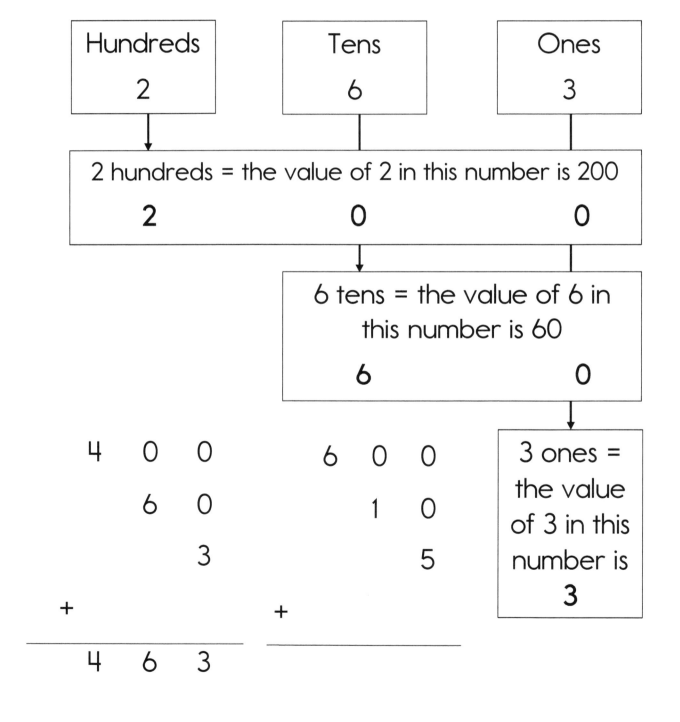

1. <u>What</u> is the value of the 4 in each of these numbers? <u>Circle</u> the right answer.

416   a) Hundreds   b) Tens   c) Ones

914   a) Hundreds   b) Tens   c) Ones

416   a) Hundreds   b) Tens   c) Ones

245   a) Hundreds   b) Tens   c) Ones

624   a) Hundreds   b) Tens   c) Ones

2. <u>Complete</u> each addition number sentence with tens and ones. The first one is done for you.

45 = 40 + 5   61 = __ + __   19 = __ + __

30 = __ + __   74 = __ + __   37 = __ + __

52 = __ + __   65 = __ + __   58 = __ + __

98 = __ + __   87 = __ + __   60 = __ + __

1st Grade Math Workbook Addition and Subtraction

1. <u>Write</u> the missing numbers.

50 + ___ = 53   ___ + ___ + 7 = 807

20 + ___ = 25   ___ + ___ + 8 = 328

90 + ___ = 96   ___ + ___ + 2 = 972

60 + ___ = 62   ___ + ___ + 0 = 430

10 + ___ = 19   ___ + ___ + 4 = 244

40 + ___ = 48   ___ + ___ + 5 = 595

2. <u>Circle</u> the correct answer.

I have a series of numbers: 0, 2, 4, 6, ___. <u>What</u> is the next number?

a) 10   b) 8   c) 12   d) 7

1. Circle the correct answer.

I have a series of numbers: 1, 2, 4, ___. What is the next number?

   a) 7   b) 8   c) 9   d) 10

2. Color the 2nd frog green. Color the 3rd frog brown. Color the 5th frog black. Color the 6th frog grey.

3. Write the missing numbers.

___ + ___ + ___ = 468   ___ + ___ + ___ = 509

___ + ___ + ___ = 135   ___ + ___ + ___ = 683

___ + ___ + ___ = 245   ___ + ___ + ___ = 248

___ + ___ + ___ = 621   ___ + ___ + ___ = 791

1. <u>Subtract.</u>

```
  146      138      157      134      159      198
-  25    -  15    -  46    -  12    -  47    -  71
─────    ─────    ─────    ─────    ─────    ─────
  121
```

```
  127      119      137      126      138      165
-  14    -  15    -  15    -  24    -  31    -  41
─────    ─────    ─────    ─────    ─────    ─────
```

2. <u>Complete</u> each pair of number bonds.

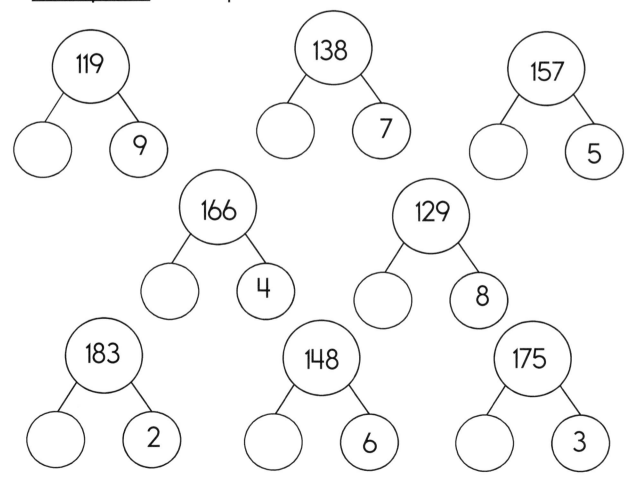

1. Complements to 100. <u>Circle</u> the missing numbers from the choice box to make the equations true.

70 + ___ = 100   a) 35   b) 30   c) 10

50 + ___ = 100   a) 40   b) 100   c) 50

90 + ___ = 100   a) 10   b) 20   c) 1

2. <u>Subtract</u>.

```
  3 5        7 2        5 4        6 8        8 7
-   5      -   2      -   4      -   8      -   7
  ─────      ─────      ─────      ─────      ─────
  3 0        _ _        _ _        _ _        _ _


  3 5        7 7        6 4        8 2        4 9
-   4      -   2      -   3      -   1      -   5
  ─────      ─────      ─────      ─────      ─────
  _ _        _ _        _ _        _ _        _ _


  6 5        9 3        5 7        3 9        8 1
- 4 0      - 2 0      - 1 0      - 2 0      - 7 0
  ─────      ─────      ─────      ─────      ─────
  _ _        _ _        _ _        _ _        _ _
```

3. The sum of the two 2-digit numbers is 50 . Their difference is 30 . What are these 2-digit numbers?

Answer: __ __ and __ __.

__ __ + __ __ = __ __   __ __ - __ __ = __ __

1. Read.

I often need to know if a number is the same as, smaller than, or larger than another number. My teacher calls this comparing numbers. Look at these candies. There are six candies in each row. My teacher says that the number in one row is **equal to** the number in the second row. 6 = 6

My sister has six candies in the top row and three candies in the bottom row. She says the number in the top row is **greater than** the number in the bottom row. 6 > 3. **6 is greater than 3.**

1. Read.

My brother has five candies in the top row and six candies in the bottom row. He says the number in the top row is **less than** the number in the second row.   5 < 6.   5 is less than 6.

2.   Circle the missing number from the choice box to make the inequality true.

5 < ___ < 9

a) 4   b) 0   c) 6

9 < ___ < 11

a) 7   b) 1   c) 10

15 < ___ < 25

a) 11   b) 20   c) 35

76 < ___ < 90

a) 89   b) 71   c) 100

1. Read.

When I round, I change a number to another number that is almost the same in value, but it is easier to work with.

For digits 0, 1, 2, and 4, we round down

For digits 5, 6, 7, 8, and 9, we round up

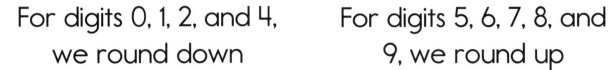

Look at 5<u>2</u>. We look at the ones digit. It's 2.

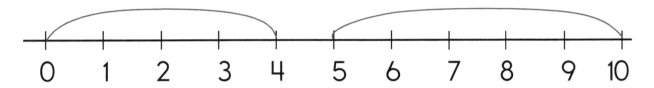

We round down to 50.

Now, look at 5<u>8</u>. The ones digit is 8.

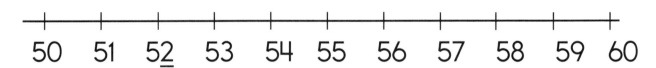

So we round UP to 60.

1. <u>Which</u> is more? <u>Compare</u> the numbers using ">," "<," or "=."

3 ones ____ 2 tens   10 ones ____ 1 ten

3 tens ____ 8 ones   2 tens ____ 1 ten

4 tens 2 ones _____ 41 ones

1 ten 2 ones _____ 1 ten 20 ones

2 tens and 3 ones _____ 3 tens and 2 ones

2. <u>Round</u> each number to the nearest 10. <u>Look</u> at the next digit to the right. If it is 0, 1, 2, 3, or 4, then ROUND DOWN, if it is 5, 6, 7, 8, 9, then ROUND UP.

<u>6</u> _____   <u>8</u> _____   <u>5</u> _____

1<u>3</u> _____   1<u>9</u> _____   1<u>6</u> _____

1<u>1</u> _____   1<u>7</u> _____   1<u>9</u> _____

2<u>2</u> _____   2<u>5</u> _____   2<u>3</u> _____

2<u>8</u> _____   2<u>4</u> _____   2<u>9</u> _____

1. Which is more? Write the missing numbers to make the comparison true.

12 ones > ___ ones   7 ones < ___ ten

1 ten = ___ ones   2 tens = ___ ones

5 ones > ___ ones   12 ones < ___ tens

15 ones < ___ ten 6 ones

1 ten 8 ones > 1 ten ___ ones

3 tens and 3 ones = ___ tens and 13 ones

2. Round each number to the nearest 100. Look at the next digit to the right. If it is 0, 1, 2, 3, or 4, then ROUND DOWN, if it is 5, 6, 7, 8, 9, then ROUND UP.

153 _____   208 _____   715 _____

913 _____   259 _____   246 _____

371 _____   557 _____   469 _____

622 _____   485 _____   923 _____

1. Read.

Even numbers are made of pairs.

An odd number is always 1 more or 1 less than an even number.

Even numbers end with a digit of 0, 2, 4, 6, 8.
Odd numbers end with a digit of 1, 3, 5, 7, 9.

2. Underline the even numbers.

1, 2, 3, 4, 5, 6, 7, 8, 9, 10, 11, 12, 13, 14, 15, 16

3. Circle the odd numbers.

15, 16, 17, 18, 19, 20, 21, 22, 23, 24, 25, 26

1. Read.

I have tons of candies. I need to estimate because it would take too long to count the exact number. I count 5 candies in the bottom row. There are 4 rows, so I can say there are about 5 + 5 + 5 + 5 candies, which is 20 candies.

I often don't need to count the candies exactly. If I have two bags of candies that cost the same, I will get the bag with more candies.

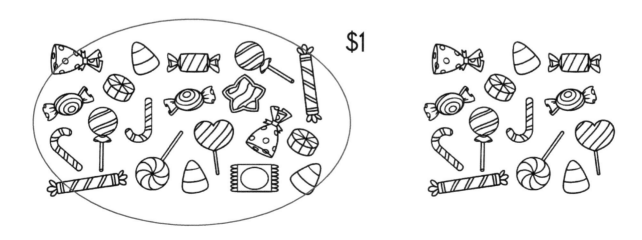

1. I need money to buy objects. <u>Write</u> the missing numbers.

3 coins: 3¢

1 + 1 + __

2 coins: 15¢

10 + __

4 coins: 40¢

10 + 10 + __ + __

3 coins: 3¢

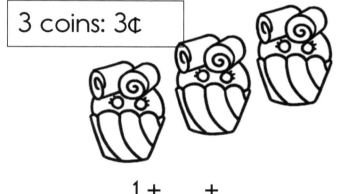

1 + __ + __

2 coins: 10¢

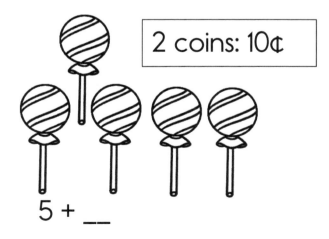

5 + __

I bought 5 lollipops for 10 cents. How much did one lollipop cost?

_____

1. I need money to buy objects. Underline{Write} the missing numbers.

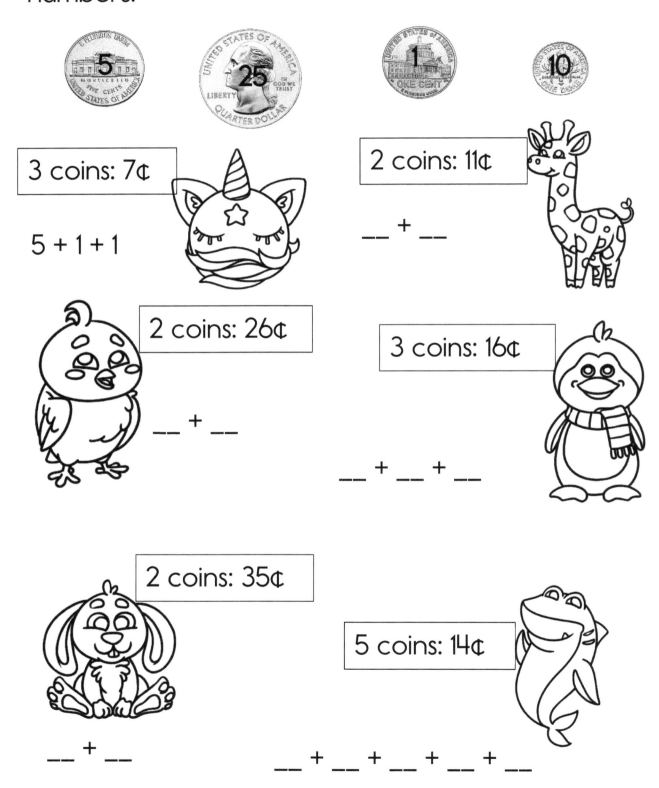

3 coins: 7¢

5 + 1 + 1

2 coins: 11¢

__ + __

2 coins: 26¢

__ + __

3 coins: 16¢

__ + __ + __

2 coins: 35¢

__ + __

5 coins: 14¢

__ + __ + __ + __ + __

1. I need money to buy objects. <u>Write</u> the missing numbers.

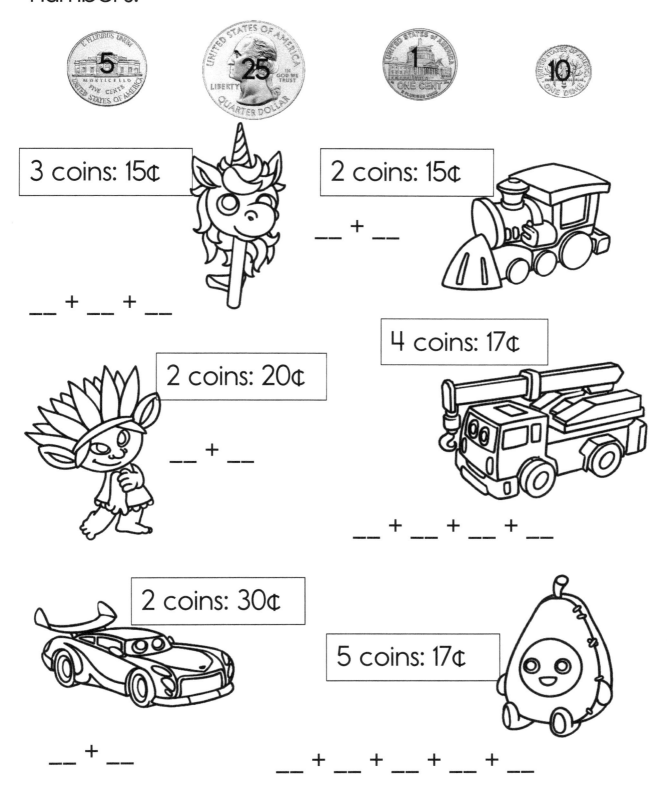

1. <u>Write</u> the missing numbers to make the scales balance.

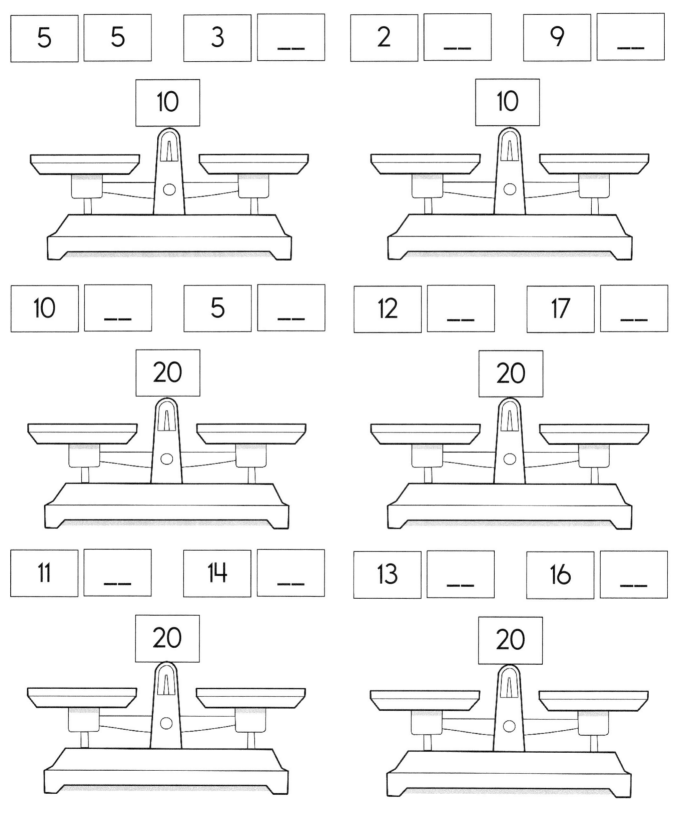

1. <u>Write</u> the missing numbers to make the scales balance.

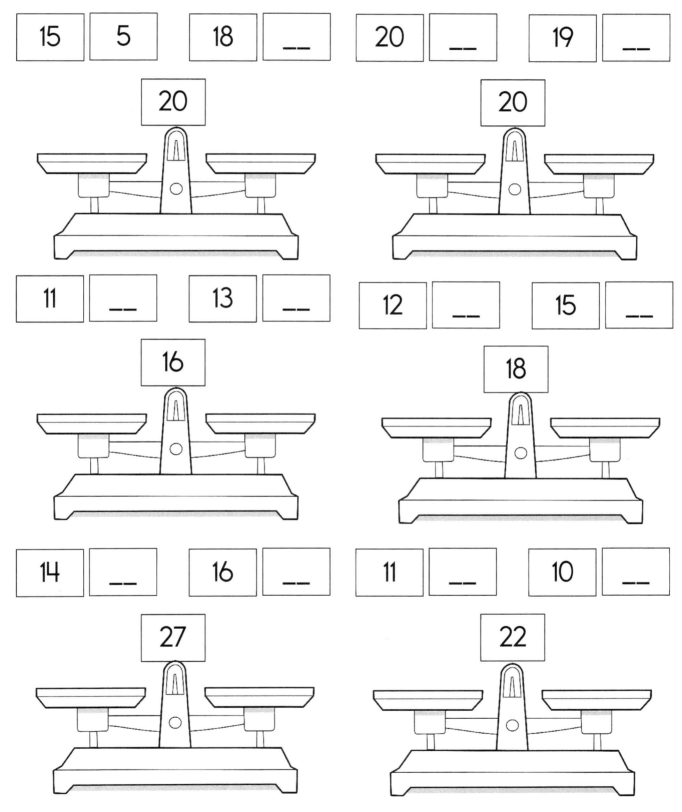

1. My Grandma baked 2 pumpkin pies. I ate a half of the total amount of pies. How many pie(s) are left?

Circle your answer:

0 ① 2 3 4 5

A half is one of two equal parts of one whole. If two pies are one whole, I could eat 1 pie which is a half. Another half is left. So, I circle 1.

I found 4 shells. My sister broke a half of the shells. Color these shells red. How many shells are left?

Circle your answer:

0 1 2 3 4 5

I got 6 cupcakes. I ate a half of them. Color the cupcakes I ate. How many cupcakes are left?

Circle your answer:

0 1 2 3 4 5

1. I got 8 candies. I ate a half of the candies. Color them red. How many candies are left?

Circle your answer:

0   1   2   3   4   5

I found 10 flowers. A half of the flowers were blooming. How many flowers were not blooming?

Circle your answer:

0   1   2   3   4   5

The pumpkin weighed 2 pounds. We ate a half of it. How many pounds are left?

Circle your answer:

0   1   2   3   4   5

My birthday cake weighed 10 pounds! My friends ate a half of the cake. How many pounds are left?

Circle your answer: 0   1   2   3   4   5

1. <u>What number</u> am I?

Half of me is 1 and double me is 4.

One whole:

A half of two equal parts of one whole is one:

Double means take twice as much or as many:

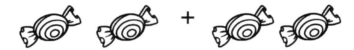

Half of me is 2 and double me is 8.

I am _____.

Half of me is 5 and double me is 20.

I am _____.

Half of me is 10 and double me is 40.

I am _____.

Half of me is 50 and double me is 200.

I am _____.

1. When you share equally between two elves, both sets of sweets and fruits have the same amount. <u>Count how many</u> for earch elf?

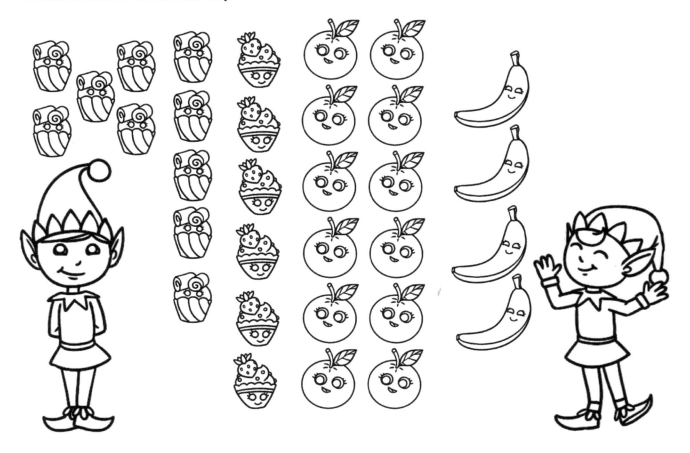

Each elf must have the same amount.

🍌	🧁	🍎	🍨
___	___	___	___

1. I am building a solid slab of rocks: the two rocks next to each other are added to get the number up above. Fill in the missing numbers.

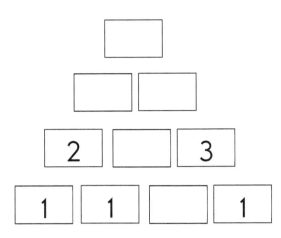

2. Write the numbers in order, from the smallest to the largest.

7, 9, 3, 1, 2, 11, 5, 12, 8, 4, 0, 10, 6

_____

3. Complements to 10. Circle the missing numbers from the choice box to make the equations true.

11 + ___ = 20   a) 8   b) 7   c) 9

5 + ___ = 20   a) 10   b) 5   c) 15

7 + ___ = 20   a) 13   b) 1   c) 14

14 + ___ = 20   a) 4   b) 16   c) 6

1. Circle the right answer.

___ + 25 = 55

d) 25
e) 30
f) 35

2. Complements to 100. Circle the missing numbers from the choice box to make the equations true.

50 + ___ = 100   a) 35   b) 50   c) 60

90 + ___ = 100   a) 10   b) 100   c) 20

20 + ___ = 100   a) 60   b) 20   c) 80

3. Circle the right answer:

I have a series of numbers: 3, 6, 9, 12, __

What is the next number?

   A) 13   B) 15   C) 18   D) 14

4. I have some numbers and signs: 1, 5, 3, +, -.

Write the equation that equals one of the answer choices.   _____

   A) 10   B) 7   C) 4   D) 9

1. Which is more? Compare the numbers using ">," "<," or "=."

13 ones _____ 2 tens   9 ones _____ 1 ten

3 tens _____ 30 ones   14 tens _____ 1 hundred

4 hundreds _____ 41 tens   2 hundreds _____ 199 ones

3 tens and 5 ones _____ 5 tens and 3 ones

2. Round each number to the nearest 10. Look at the next digit to the right. If it is 0, 1, 2, 3, or 4 then ROUND DOWN, if it is 5, 6, 7, 8, 9 then, ROUND UP.

23 _____   44 _____   26 _____   51 _____
87 _____   39 _____   72 _____   65 _____

3. What number am I?

Half of me is 20 and double me is 80. _____

Half of me is 10 and double me is 40. _____

Half of me is 25 and double me is 100. _____

Half of me is 11 and double me is 44. _____

1. <u>Solve</u> the problem:

1047: the sum of the ones and hundreds is _____.

A) 5
B) 8
C) 7

2. I start at 0 and count on in twos. Will I say 11?

_____

Why? _____

I start at 0 and count on in twos. Will I say 16?

_____

Why? _____

3. <u>Find</u> the value.

853:

The sum of the ones and tens is _____.

The difference between the hundreds and tens is _____.

The difference between the hundreds and ones is _____.

1. Solve the problems:

I had 8 candies. I gave 4 of them to my sister. How many candies has I left? _____

There are 10 kids at a playground. I counted 7 boys. How many girls are there?

_____

My brother bought 11 chocolate cupcakes and 5 vanilla cupcakes. How many cupcakes did he buy in all?

_____

I found 6 easter eggs. My sister found 4 more Easter eggs than I did. My brother found 7 less Easter eggs than my sister. How many Easter eggs did my brother find?

_____

1. <u>Solve</u> the problems:

I had 15 cupcakes. I ate some cupcakes, and I had 12 cupcakes left. <u>How many cupcakes</u> did I eat?

_____

My brother has 18 trucks and race cars. 3 of them are trucks. <u>How many race cars</u> does he have?

_____

My sister saw 11 butterflies. My brother saw 4 butterflies more than my sister. <u>How many butterflies</u> did they see altogether?

_____

1. Solve the problems:

I had some cupcakes. I ate 3 cupcakes and I gave 4 cupcakes to my friend. I have 11 cupcakes left. How many cupcakes did I have at first?

_____

I had 5 yellow balloons and 5 more red balloons than yellow balloons. My friend had 8 more ballons than I had. How many balloons did my friend have?

_____

There are 6 oranges in a basket. My mother puts 10 small pears and 3 bananas into the basket. How many fruits are there in the basket altogether?

_____

Hint: Write ones under ones. Carry over when the sum is 10 or more. Find out how many more you need to add to a greater number to get a ten.

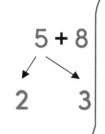

5 + 8
2   3

Decompose a smaller number. It's 5.

5 is 2+3 and I need to add two more to 8 to get 10.

```
  ones
    5
+   8
  ───
```

In columns I add one more row to write the numbers that were carried over, right?

tens ones
  –
       5
+   8
  ─────
  –   –

Step 1: I need one more row above 5.
Step 2: First, I add ones: 5+8=13.

Write the 3 in one's place. Carry 1 ten with the tens. Write the 1 in ten's place.

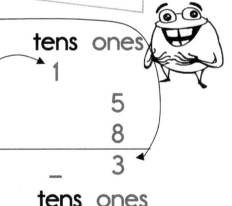

tens ones
  1
       5
       8
  ─────
       3

tens ones
  1
       5
+   8
  ─────
  1   3

Step 3: Add tens: 1+0+0=1. Rewrite the 1 in ten's place.

1. <u>Add.</u>

```
  -9      -9      -9      -9      -9
 + 7     + 6     + 4     + 5     + 8
 ----    ----    ----    ----    ----

  -9      -9      -9      -8      -8
 + 2     + 1     + 3     + 6     + 8
 ----    ----    ----    ----    ----

  -8      -8      -8      -8      -8
 + 7     + 9     + 4     + 2     + 5
 ----    ----    ----    ----    ----
```

1. <u>Add.</u>

```
   8       7       7       7       7
+  3    +  6    +  4    +  5    +  8
____    ____    ____    ____    ____


   7       7       7       6       6
+  3    +  7    +  9    +  6    +  8
____    ____    ____    ____    ____


   6       6       6       6       5
+  7    +  9    +  4    +  5    +  5
____    ____    ____    ____    ____
```

2.

Circle the right answer:

I have a series of numbers:
0, 2, 2, 4, 6, __

What is the next number?

A 6   C 10

B 12   D 9

1. <u>Add.</u>

$$\begin{array}{r}5\\+\ 9\\\hline\end{array} \quad \begin{array}{r}5\\+\ 6\\\hline\end{array} \quad \begin{array}{r}5\\+\ 7\\\hline\end{array} \quad \begin{array}{r}5\\+\ 8\\\hline\end{array} \quad \begin{array}{r}4\\+\ 8\\\hline\end{array}$$

$$\begin{array}{r}4\\+\ 6\\\hline\end{array} \quad \begin{array}{r}4\\+\ 7\\\hline\end{array} \quad \begin{array}{r}4\\+\ 9\\\hline\end{array} \quad \begin{array}{r}3\\+\ 7\\\hline\end{array} \quad \begin{array}{r}3\\+\ 8\\\hline\end{array}$$

$$\begin{array}{r}3\\+\ 9\\\hline\end{array} \quad \begin{array}{r}2\\+\ 9\\\hline\end{array} \quad \begin{array}{r}8\\+\ 9\\\hline\end{array} \quad \begin{array}{r}1\\+\ 9\\\hline\end{array} \quad \begin{array}{r}5\\+\ 5\\\hline\end{array}$$

2.

Circle the right answer:

I have a series of numbers: 19, 12, 7, 4, __

What is the next number?

A   5   C   2
B   3   D   1

1.

Circle the right answer:

I have a series of numbers: 3, 5, 9, __

What is the next number?

A   11   C   15
B   13   D   9

Hint: Write the smaller number under the larger number; Ones under ones, tens under tens; Subtract ones, then tens.

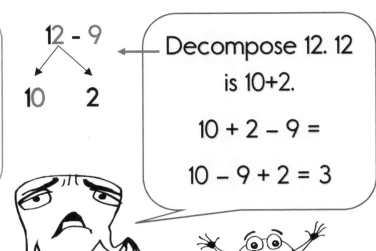

Decompose 12. 12 is 10+2.

10 + 2 − 9 =

10 − 9 + 2 = 3

In the columns we subtract 9 ones out of 2 ones.

Step 1: If I subtract in columns, I need one more row above 12.

Step 2: Borrow 1 ten out of the tens since 12=10+2.

Write 12 above 2 in one's place since 10 ones and 2 ones are 12.

```
tens ones
       12
  1     2
 −      9
```

Step 3: leave 0 above 1 in ten's place. Cross out 1 and 2 to avoid mistakes.

Step 3: Subtract 9 ones from 12 ones: 12−9=3. Hint: Do NOT rewrite 0 in ten's place.

```
tens ones
  0    12
  1̶     2̶
 −      9
        3
```

1. Subtract.

```
  0 14
  1̶ 4      1 5      1 2      1 5      1 6
-   9    -   8    -   6    -   7    -   8
_____  _____  _____  _____  _____


  1 4      1 7      1 4      1 6      1 3
-   7    -   8    -   8    -   7    -   9
_____  _____  _____  _____  _____


  1 8      1 2      1 5      1 3      1 5
-   9    -   8    -   9    -   9    -   8
_____  _____  _____  _____  _____
```

2. Solve the problem and write the missing number:

10 candies   =   5 cupcakes

2 candies = __ cupcake(s)

Answer: _____.

1. <u>Add.</u>   Score ___/15   Time __:__

```
  ‾4      ‾5      ‾9      ‾5      ‾6
+  9    +  8    +  7    +  5    +  8
────    ────    ────    ────    ────

  ‾4      ‾7      ‾4      ‾6      ‾9
+  7    +  7    +  8    +  7    +  3
────    ────    ────    ────    ────

  ‾8      ‾2      ‾5      ‾3      ‾5
+  9    +  8    +  9    +  9    +  8
────    ────    ────    ────    ────
```

2. I have some numbers and signs: 2, 3, 3, +, -.
<u>Write</u> the equation that equals one of the answer choices.

A 1   C 4
B 5   D 7   _____

1. <u>Subtract.</u>   Score ___/15   Time __:__

```
  16      12      14      11      15
-  9    -  8    -  7    -  5    -  8
____    ____    ____    ____    ____

  11      15      11      13      15
-  7    -  7    -  8    -  7    -  9
____    ____    ____    ____    ____

  12      13      14      17      14
-  9    -  8    -  9    -  9    -  8
____    ____    ____    ____    ____
```

2.

I have some numbers and signs: 4, 5, 7, +, -.

<u>Write</u> the equation that equals one of the answer choices.

A 4   C 10
B 6   D 9   _____

1. Add.   Score ___/15   Time __:__

```
   9      6      7      8      5
+  9   +  6   +  7   +  8   +  5
_____  _____  _____  _____  _____

   4      3      4      5      4
+  9   +  7   +  8   +  7   +  7
_____  _____  _____  _____  _____

   6      3      8      5      3
+  9   +  9   +  9   +  9   +  9
_____  _____  _____  _____  _____
```

2. I have some numbers and signs: 12, 7, -.

Write the equation that equals one of the answer choices.

A 6   C 7
B 4   D 5   _____

1. Solve the problem:

9728: the difference of the thousands and the tens is _____.

Answer:

_____

1. <u>Subtract.</u>   Score ___/15   Time __:__

```
  16      12      14      11      15
-  7    -  3    -  5    -  4    -  6
____    ____    ____    ____    ____

  11      15      11      13      15
-  3    -  8    -  7    -  6    -  8
____    ____    ____    ____    ____

  12      13      14      17      14
-  6    -  5    -  7    -  8    -  9
____    ____    ____    ____    ____
```

2.

I have some numbers and signs: 17, 8, -. <u>Write</u> the equation that equals one of the answer choices.

A   8   C   9
B   11   D   7   _____

1. Add.   Score ___/15   Time __:__

```
  6      6      6      6      6
+ 9    + 6    + 7    + 8    + 4
____   ____   ____   ____   ____

  8      8      8      8      8
+ 6    + 2    + 5    + 7    + 8
____   ____   ____   ____   ____

  7      3      7      7      7
+ 9    + 9    + 3    + 6    + 5
____   ____   ____   ____   ____
```

2.

I have some numbers and signs: 9, 4, +.

Write the equation that equals one of the answer choices.

A 12   C 14
B 13   D 11   _____

1. Subtract.   Score ___/15   Time __:__

```
 1 1     1 1     1 1     1 1     1 1
-  7    -  3    -  5    -  4    -  6
----    ----    ----    ----    ----
```

```
 1 2     1 2     1 2     1 2     1 2
-  3    -  8    -  7    -  6    -  8
----    ----    ----    ----    ----
```

```
 1 3     1 3     1 3     1 3     1 3
-  6    -  5    -  7    -  8    -  9
----    ----    ----    ----    ----
```

2.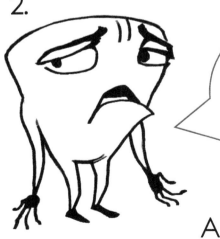

I have some numbers and signs: 3, 3, 5, +, +.

<u>Write</u> the equation that equals one of the answer choices.

A   10   C   13
B   12   D   11   _____

1. <u>Subtract.</u> Score ___/15   Time __:__

```
  1 6     1 2     1 4     1 1     1 5
-   8   -   5   -   9   -   2   -   7
-----   -----   -----   -----   -----


  1 1     1 5     1 1     1 3     1 5
-   6   -   9   -   4   -   8   -   8
-----   -----   -----   -----   -----


  1 2     1 3     1 4     1 7     1 4
-   4   -   6   -   6   -   9   -   6
-----   -----   -----   -----   -----
```

2.

Circle the right answer:

I have a series of numbers: 3, 6, 5, 8, 7, ___.

What is the next number?

A 10   C 11
B 12   D 9

1. <u>Add.</u>   Score ___/15   Time __:__

$\begin{array}{r}7\\+\ 9\\\hline\end{array}$ $\begin{array}{r}6\\+\ 8\\\hline\end{array}$ $\begin{array}{r}4\\+\ 7\\\hline\end{array}$ $\begin{array}{r}6\\+\ 9\\\hline\end{array}$ $\begin{array}{r}8\\+\ 4\\\hline\end{array}$

$\begin{array}{r}7\\+\ 6\\\hline\end{array}$ $\begin{array}{r}8\\+\ 5\\\hline\end{array}$ $\begin{array}{r}9\\+\ 5\\\hline\end{array}$ $\begin{array}{r}8\\+\ 9\\\hline\end{array}$ $\begin{array}{r}7\\+\ 8\\\hline\end{array}$

$\begin{array}{r}7\\+\ 3\\\hline\end{array}$ $\begin{array}{r}2\\+\ 9\\\hline\end{array}$ $\begin{array}{r}7\\+\ 5\\\hline\end{array}$ $\begin{array}{r}4\\+\ 6\\\hline\end{array}$ $\begin{array}{r}7\\+\ 7\\\hline\end{array}$

2.

Circle the right answer:
I have a series of numbers: 5, 11, 7, 13, 9, __.

A 19   C 15
B 13   D 17

1. Add.   Score ___/15   Time __:__

```
  9     4     7     6     8
+ 9   + 8   + 7   + 5   + 8
____  ____  ____  ____  ____

  9     5     9     8     9
+ 6   + 5   + 7   + 7   + 8
____  ____  ____  ____  ____

  7     2     9     8     7
+ 5   + 8   + 5   + 6   + 8
____  ____  ____  ____  ____
```

2. I have some numbers and signs: 6, 7, 11, +, -.

Write the equation that equals one of the answer choices.

A 13   C 4
B 2   D 16   _____

1. Subtract.   Score ___/15   Time __:__

```
  16      12      14      11      15
-  8    -  6    -  7    -  5    -  7
____    ____    ____    ____    ____

  11      17      12      13      17
-  4    -  8    -  7    -  8    -  8
____    ____    ____    ____    ____

  12      13      14      17      14
-  3    -  9    -  8    -  9    -  5
____    ____    ____    ____    ____
```

2.

I have some numbers and signs: 4, 6, 8, +, -.

Write the equation that equals one of the answer choices.

A 9   C 6
B 7   D 12   _____

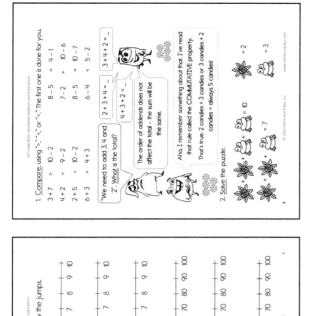
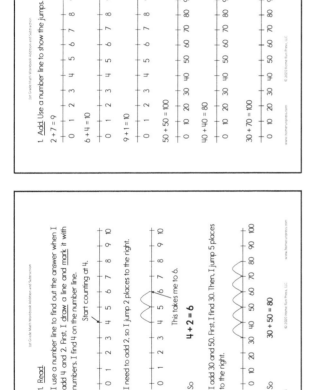
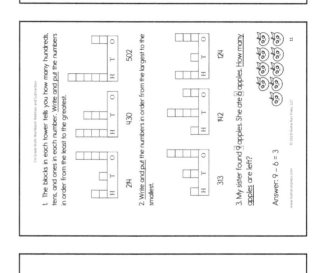
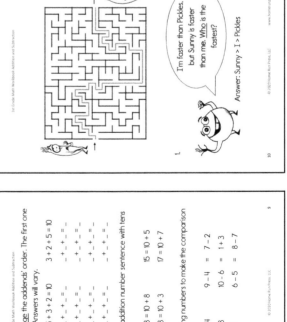
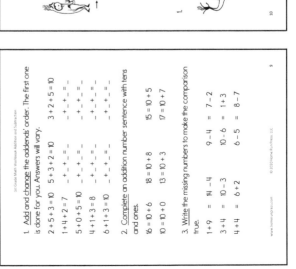

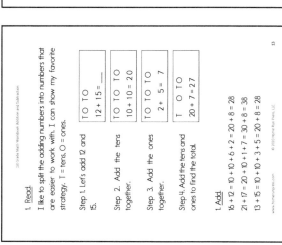
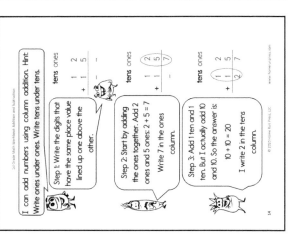
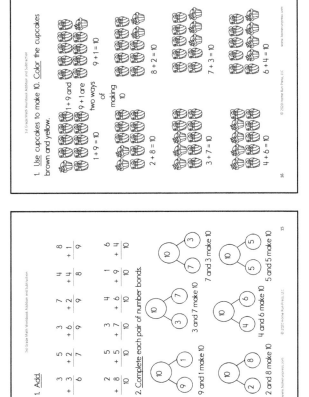
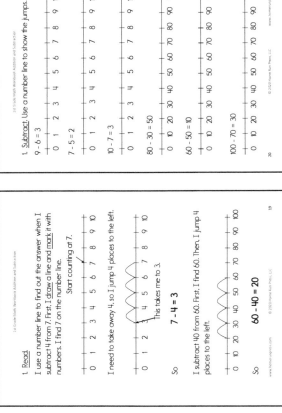
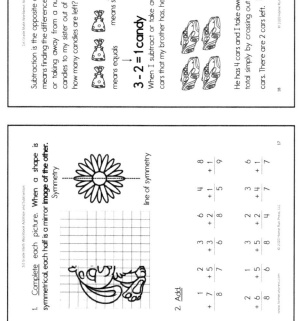
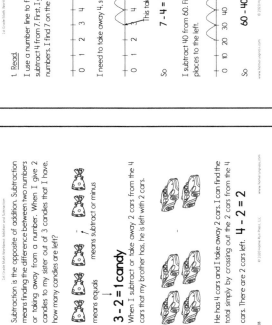

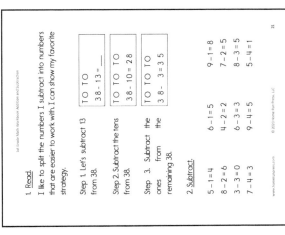
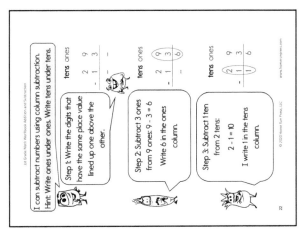
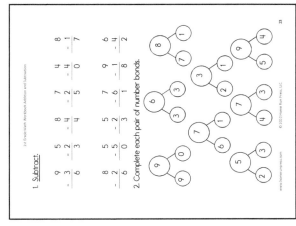
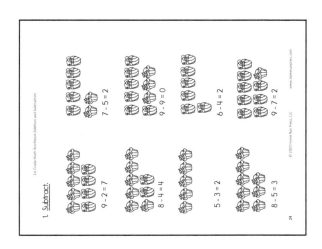
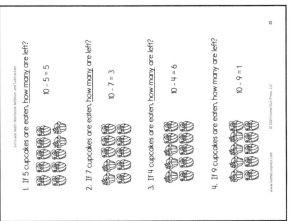
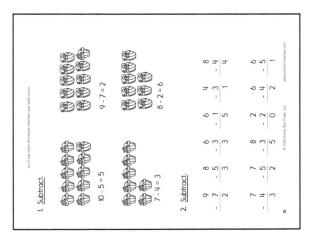
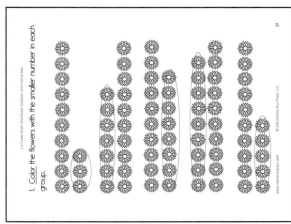
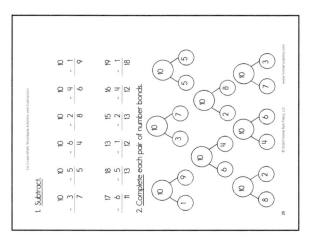

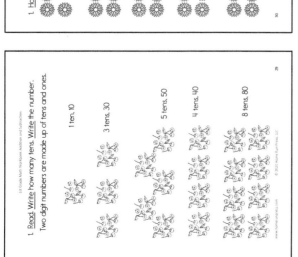
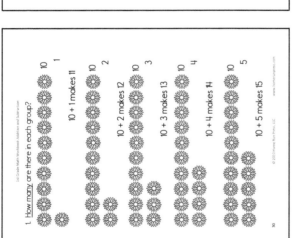
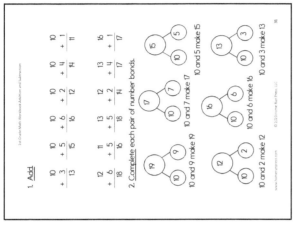
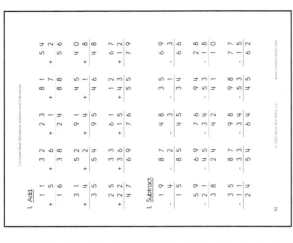
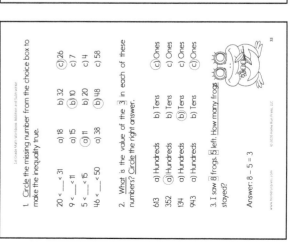
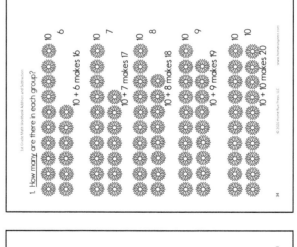
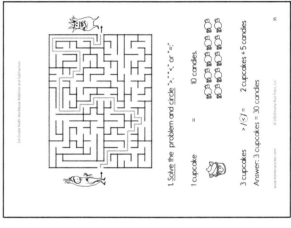
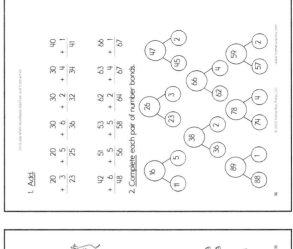

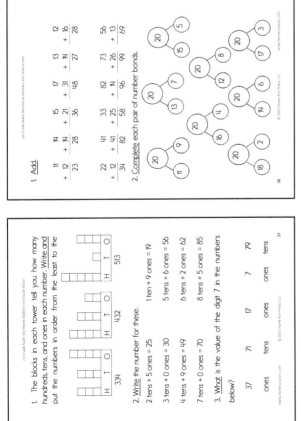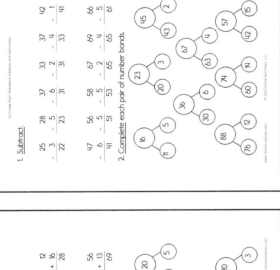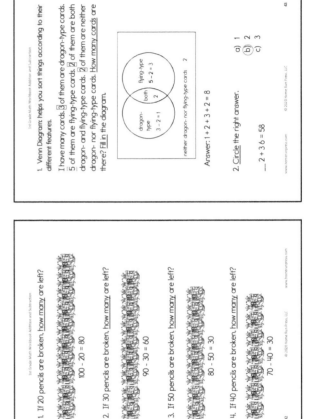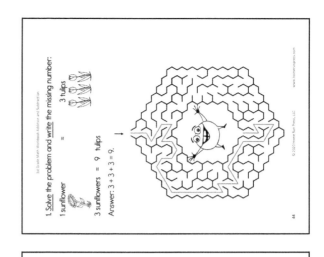

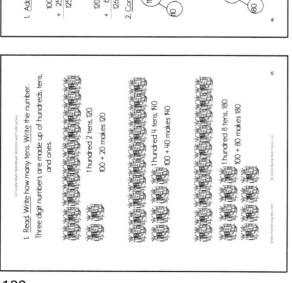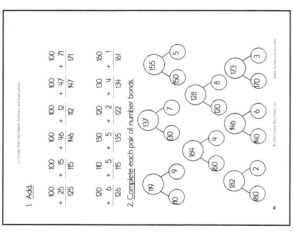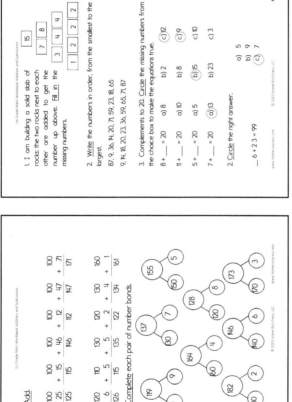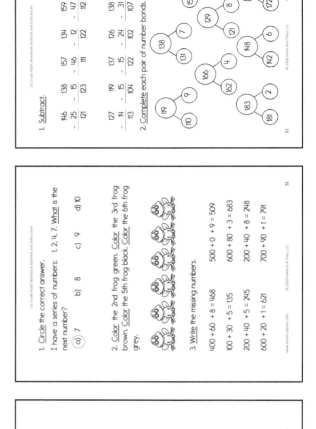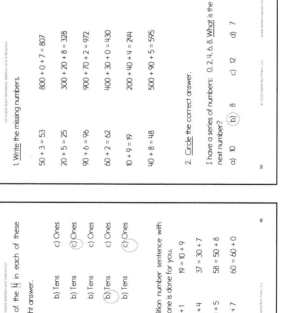

1. Complements to 100. Circle the missing numbers from the choice box to make the equations true.

70 + ___ = 100   a) 35   b) 30   c) 10
50 + ___ = 100   a) 40   b) 100   (c) 50
90 + ___ = 100   a) (10) b) 20   c) 1

2. Subtract.

```
  3 5        5 4        6 8           8 7
- 3 0      - 5 0      - 8          - 8 0
-----      -----      -----         -----
              4           60           7
```

```
  3 5        7 7        6 4        8 2          4 9
- 4        - 2        - 3        - 1          - 5
-----      -----      -----      -----        -----
  31         75         61         81           44
```

```
  6 5        9 3        5 7        3 9          8 1
- 2 0      - 1 0      - 1 0      - 2 0        - 7 0
-----      -----      -----      -----        -----
  45         83         47         19           11
```

3. The sum of the two 2-digit numbers is 50. Their difference is 30. What are these 2-digit numbers?

Answer: 40 and 10.

40 + 10 = 50   40 - 10 = 30

---

1. Read.

I often need to know if a number is the same as, smaller than, or larger than another number. My teacher calls this comparing numbers. Look at these candies. There are six candies in each row. My teacher says that the number in one row is equal to the number in the second row. **6 = 6**

My sister has six candies in the top row and three candies in the bottom row. She says the number in the top row is greater than the number in the bottom row. **6 > 3. 6 is greater than 3.**

---

1. Read.

My brother has five candies in the top row and six candies in the bottom row. He says the number in the top row is **less than** the number in the second row.   5 < 6.   **5 is less than 6.**

2. Circle the missing number from the choice box to make the inequality true.

5 < ___ < 9

a) 4   b) 0   (c) 6
a) 9   b) 1   (c) 10
a) 7   b) 1   c) 10

15 < ___ < 25
a) 11   (b) 20   c) 35
76 < ___ < 90
a) 89   (b) 71   c) 100

---

1. Read.

When I round, I change a number to another number that is almost the same in value, but it is easier to work with.

For digits 0, 1, 2, and 4,   For digits 5, 6, 7, 8, and
we round down   9, we round up

0 1 2 3 4 5 6 7 8 9 10

Look at 52. We look at the ones digit. It's 2.

50 51 52 53 54 55 56 57 58 59 60

We round down to 50.

Now, look at 58. The ones digit is 8.

50 51 52 53 54 55 56 57 58 59 60

So we round UP to 60.

---

1. Which is more? Compare the numbers using ">," "<," or "=".

3 ones < 2 tens   10 ones = 1 ten   5 10
3 tens > 8 ones   2 tens > 1 ten
4 tens 2 ones > 41 ones
1 ten 2 ones < 1 ten 20 ones
2 tens and 3 ones < 3 tens and 2 ones

2. Round each number to the nearest 10. Look at the next digit to the right. If it is 0, 1, 2, 3, or 4, then ROUND DOWN. If it is 5, 6, 7, 8, 9, then ROUND UP.

6 10   8 10   19 20   5 10
13 10   17 20   19 20   16 20
22 20   25 30   23 20
28 30   24 20   29 30

---

1. Which is more? Write the missing numbers to make the comparison true. Answers may vary.

12 ones > 8 ones   7 ones < 1 ten
1 ten = 10 ones   2 tens = 20 ones
5 ones > 3 ones   12 ones < 2 tens
15 ones < 1 ten 6 ones
1 ten 8 ones > 1 ten 5 ones
3 tens and 3 ones = 2 tens and 13 ones

2. Round each number to the nearest 100. Look at the next digit to the right. If it is 0, 1, 2, 3, or 4, then ROUND DOWN. If it is 5, 6, 7, 8, 9, then ROUND UP.

153 200   208 200   200   715 700
913 900   259 300   246   200
371 400   557 600   469   500
622 600   485 500   923   900

---

1. Read.

Even numbers are made of pairs.

An odd number is always 1 more or 1 less than an even number.

Even numbers end with a digit of 0, 2, 4, 6, 8.
Odd numbers end with a digit of 1, 3, 5, 7, 9.

2. Underline the even numbers.

1, 2, 3, 4, 5, 6, 7, 8, 9, 10, 11, 12, 13, 14, 15, 16

3. Circle the odd numbers.

(15), 16, (17), 18, (19), 20, (21), 22, (23), 24, (25), 26

---

1. Read.

I have tons of candies. I need to estimate because it would take too long to count the exact number. I count 5 candies in the bottom row. There are 4 rows, so I can say there are about 5 + 5 + 5 + 5 candies, which is 20 candies.

I often don't need to count the candies exactly. If I have two bags of candies that cost the same, I will get the bag with more candies.

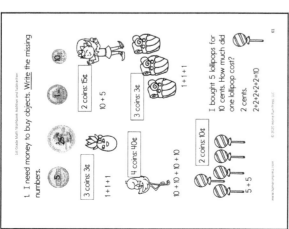
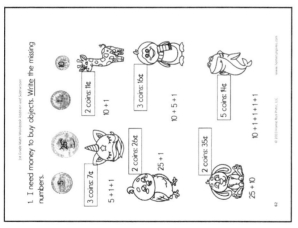
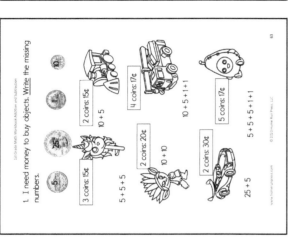
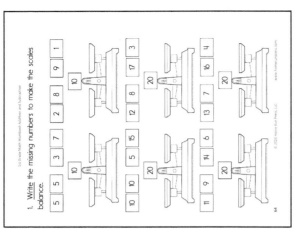
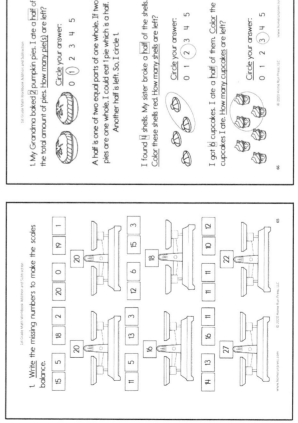
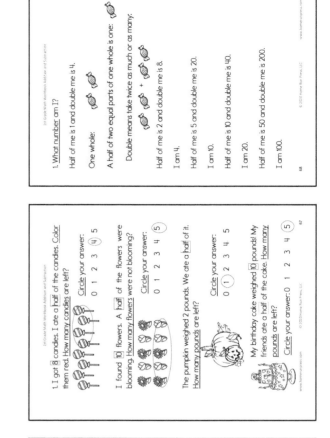

Page 69

1. When you share equally between two elves, both sets of sweets and fruits have the same amount. Count how many for each elf?

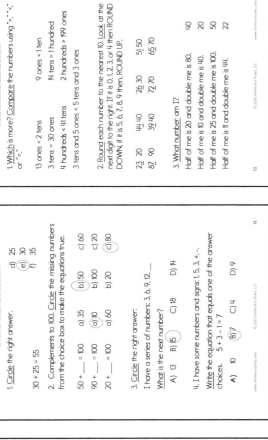

Each elf must have the same amount.

2	5	6
		3

Page 70

1. I am building a solid slab of rocks: the two rocks next to each other are added to get the number up above. Fill in the missing numbers.

```
      11
    5   6
  2   3   3
1   1   2   1
```

2. Write the numbers in order, from the smallest to the largest.
7, 9, 3, 1, 2, 11, 5, 12, 8, 4, 0, 10, 6
0, 1, 2, 3, 4, 5, 6, 7, 8, 9, 10, 11, 12

3. Complements to 10. Circle the missing numbers from the choice box to make the equations true.

11 + 9 = 20   a) 8   b) 7   c)9
5 + 15 = 20   a) 10   b) 5   c)15
7 + 13 = 20   a) 13   b) 1   c) 14
14 + 6 = 20   a) 4   b) 16   c) 6

Page 71

1. Circle the right answer.
30 + 25 = 55   d) 25
                e) 30
                f) 35

2. Complements to 100. Circle the missing numbers from the choice box to make the equations true.
50 + ___ = 100   a) 35   b) 50   c) 60
90 + ___ = 100   a) 10   b) 100   c) 20
20 + ___ = 100   a) 60   b) 20   c) 80

3. Circle the right answer:
I have a series of numbers: 3, 6, 9, 12, ___
What is the next number?
A) 13   B) 15   C) 18   D) 14

4. I have some numbers and signs: 1, 5, 3, +, -.
Write the equation that equals one of the answer choices.   5 + 3 - 1 = 7
A) 10   B) 7   C) 4   D) 9

Page 72

1. Which is more? Compare the numbers using ">", "<", or "=".
13 ones < 2 tens   9 ones < 1 ten
3 tens = 30 ones   14 tens > 1 hundred
4 hundreds < 41 tens   2 hundreds > 199 ones
3 tens and 5 ones < 5 tens and 3 ones

2. Round each number to the nearest 10. Look at the next digit to the right. If it is 0, 1, 2, 3, or 4 then ROUND DOWN. If it is 5, 6, 7, 8, 9 then, ROUND UP.

23 20   26 30   51 50
87 90   39 40   72 70   65 70

3. What number am I?
Half of me is 20 and double me is 80.   40
Half of me is 10 and double me is 40.   20
Half of me is 25 and double me is 100.   50
Half of me is 11 and double me is 44.   22

Page 73

1. Solve the problem:
1047: the sum of the ones and hundreds is 7 + 0 = 7.
A) 5   B) 8   C) 7

2. I start at 0 and count on in twos. Will I say 11? 0, 2, 4, 6, 8, 10, 12
Why? 11 is an odd number.

I start at 0 and count on in twos. Will I say 16? 0, 2, 4, 6, 8, 10, 12, 14, 16
Why? 16 is an even number.

3. Find the value.
853:
The sum of the ones and tens is 3 + 5 = 8.
The difference between the hundreds and tens is 8 − 5 = 3.
The difference between the hundreds and ones is 8 − 3 = 5.

Page 74

1. Solve the problems:

I had 8 candies. I gave 4 of them to my sister. How many candies has I left?   8 − 4 = 4 (candies)

There are 10 kids at a playground. I counted 7 boys. How many girls are there?   3 girls.
10 − 7 = 3 (girls)

My brother bought 11 chocolate cupcakes and 5 vanilla cupcakes. How many cupcakes did he buy in all?   16 cupcakes.
11 + 5 = 16 (cupcakes)

I found 6 easter eggs. My sister found 4 more Easter eggs than I did. My brother found 7 less Easter eggs than my sister. How many Easter eggs did my brother find?   3 Easter eggs.
6 + 4 = 10 (sister)   10 − 7 = 3 (brother)

Page 75

1. Solve the problems:

I had 15 cupcakes. I ate some cupcakes, and I had 12 cupcakes left. How many cupcakes did I eat?
3 cupcakes.   15 − 12 = 3

My brother has 18 trucks and race cars. 3 of them are trucks. How many race cars does he have?
15 race cars.   18 − 3 = 15

My sister saw 11 butterflies. My brother saw 4 butterflies more than my sister. How many butterflies did they see altogether?
26 butterflies.   11 + 4 + 11 = 26

Page 76

1. Solve the problems:

I had some cupcakes. I ate 3 cupcakes and I gave 4 cupcakes to my friend. I have 11 cupcakes left. How many cupcakes did I have at first?
18 cupcakes: 3 + 4 + 11 = 18

I had 5 yellow balloons and 5 more red balloons than yellow balloons. My friend had 8 more balloons than I had. How many balloons did my friend have?
18 balloons: 5 + 5 + 8 = 18

There are 8 oranges in a basket. My mother puts 10 small pears and 3 bananas into the basket. How many fruits are there in the basket altogether?
19 fruits: 6 + 10 + 3 = 19

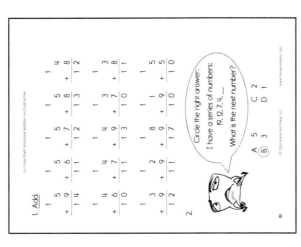

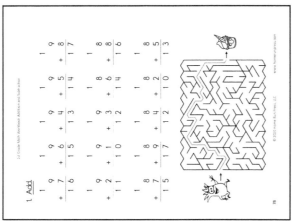
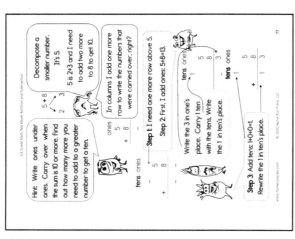
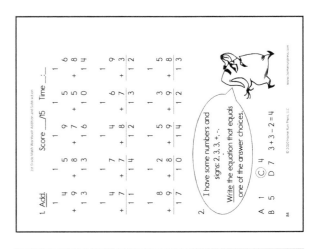
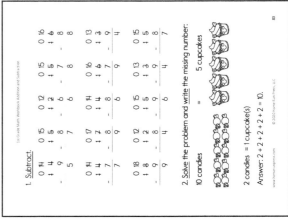
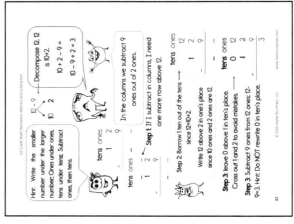
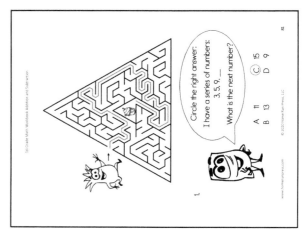

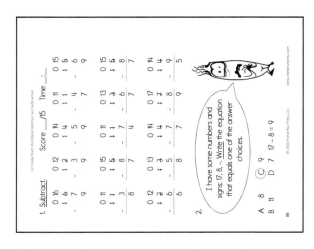

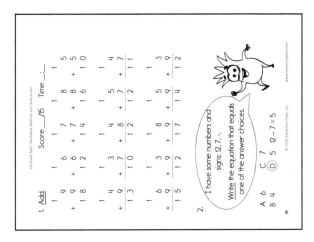
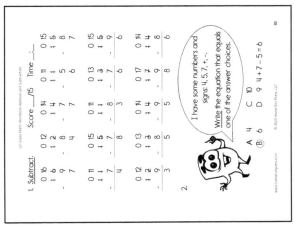
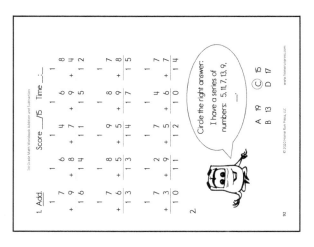
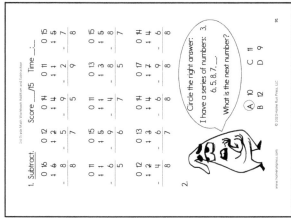
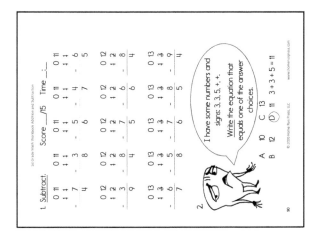
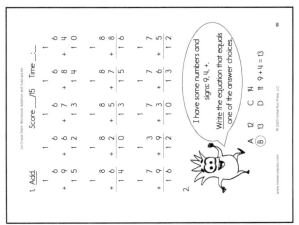

Page 93

1. Add.

¹9	¹4	7	6	⁺8
+9	+8	+7	+5	+8
18	12	14	11	16

¹9	5	¹9	¹8	¹9
+6	+5	+7	+7	+8
15	10	16	15	17

7	¹2	9	¹9	7
+5	+8	+5	+6	+8
14	10	14	15	15

2. I have some numbers and signs: 6, 7, 11, +, −. Write the equation that equals one of the answer choices.

A 13   C 4
B 2 (circled)   D 16 $6 + 7 - 11 = 2$

Page 94

1. Subtract.

⁰1̸6	⁰1̸2	⁰1̸4	⁰1̸1	⁰1̸5
−8	−6	−7	−6	−7
8	6	7	5	8

⁰1̸1	⁰1̸7	⁰1̸2	⁰1̸3	⁰1̸7
−4	−8	−7	−8	−8
7	9	5	5	9

⁰1̸2	⁰1̸3	⁰1̸4	⁰1̸7	⁰1̸4
−3	−9	−8	−9	−5
9	4	6	8	9

2. I have some numbers and signs: 4, 6, 8, +, −. Write the equation that equals one of the answer choices.

A 9   C 6 (circled)
B 7   D 12 $4 + 8 - 6 = 6$

Printed in the USA
CPSIA information can be obtained
at www.ICGtesting.com
LVHW082228111124
796359LV00044B/1589